HAZARDOUS AIR POLLUTANT HANDBOOK

Measurements,
Properties,
and Fate
in Ambient Air

Chester W. Spicer
Sydney M. Gordon
Michael W. Holdren
Thomas J. Kelly
R. Mukund

CRC Press
Taylor & Francis Group
Boca Raton London New York

CRC Press is an imprint of the
Taylor & Francis Group, an **informa** business

First published 2002 by Lewis Publishers

Published 2020 by CRC Press
Taylor & Francis Group
6000 Broken Sound Parkway NW, Suite 300
Boca Raton, FL 33487-2742

First issued in paperback 2020

© 2002 by Taylor & Francis Group, LLC
CRC Press is an imprint of Taylor & Francis Group, an Informa business

No claim to original U.S. Government works

ISBN 13: 978-0-367-57861-9 (pbk)
ISBN 13: 978-1-56670-571-4 (hbk)

**Visit the Taylor & Francis Web site at
http://www.taylorandfrancis.com**

**and the CRC Press Web site at
http://www.crcpress.com**

Library of Congress Cataloging-in-Publication Data

Hazardous air pollutant handbook : measurements, properties, and fate in ambient air /
Chester W. Spicer ... [et al].
     p. cm.
    Includes bibliographical references and index.
    ISBN 1-56670-571-1 (alk. paper)
     1. Air—Pollution--Handbooks, manuals, etc. 2. Pollutants--Handbooks, manuals, etc. I.
  Spicer, Chester W.
TD883 .H396 2002
628.5'—dc21

2002017540

Library of Congress Card Number 2002017540

Acknowledgments

The authors would like to thank the numerous people who have contributed to this effort. Battelle colleagues Al Pollack, Melinda Armbruster, and Susan Abbgy participated in the literature surveys, and Leanna House and Steve Bortnick contributed to the update of the ambient concentration survey. We are indebted to the U.S. Environmental Protection Agency and Drs. Larry Cupitt, William McClenny, and Robert Lewis for support and advice. We also wish to acknowledge the Atmospheric Science and Applied Technology Department of Battelle for providing secretarial support and other resources during the production of this handbook, and Mary Ann Roberts for assistance with manuscript preparation.

The Authors

Chester W. Spicer is a senior research leader in the Atmospheric Science and Applied Technology Department of Battelle in Columbus, OH. His academic background includes a B.A. in chemistry from Rutgers University and a Ph.D. in analytical chemistry from Pennsylvania State University. His principal interests and activities include the elucidation of atmospheric chemical transformations, studies of the distribution and fate of hazardous air pollutants, indoor and outdoor air quality and development of analytical methods for gas and aerosol measurement. He has directed numerous scientific investigations utilizing research aircraft, mobile laboratories, research houses and smog chambers.

Dr. Spicer's recent research includes studies of urban photochemical air pollution through monitoring from aircraft and skyscrapers, studies of gaseous halogen contributions to ozone depletion in the Arctic, the identification and sources of molecular halogens in the marine atmosphere, chemical transformations of oxidized nitrogen compounds in indoor environments, the sources and variability of hazardous air pollutants in urban areas, and emissions of toxic pollutants from diverse sources including aircraft engines, natural gas appliances and military munitions.

He is past chairman of the Editorial Review Board of the Air and Waste Management Association, and has served on the Editorial Board of the *Journal of Environmental Forensics*. He recently served on the National Academy of Sciences/National Research Council Panel on Atmospheric Effects of Aviation. Dr. Spicer is a member of the American Geophysical Union, the American Chemical Society, the American Association for the Advancement of Science, the Air and Waste Management Association and the International Society of Indoor Air Quality and Climate.

Sydney M. Gordon is a research leader in the Atmospheric Science and Applied Technology Department of Battelle in Columbus. His research focuses on the development and application of methods for measuring trace-level pollutants in ambient and indoor air. The characterization of pollutants in human exhaled breath is a particular area of interest for application in human exposure assessments.

At Battelle, Dr. Gordon leads a human exposure assessment group that concentrates on large multipollutant, multimedia studies, and manages a variety of programs funded by government and industrial clients. Much of his recent work has dealt with the development of new mass spectrometric techniques for use in environmental and biomedical problems. He is the author or co-author of more than 120 research publications, book chapters, and presentations.

Dr. Gordon received his D.Sc. in physical chemistry from the University of Pretoria, South Africa, in 1965. He is a member of the American Chemical Society, the International Society of Exposure Analysis, and the American Society for Mass Spectrometry. He is listed in *American Men and Women of Science*, *Who's Who in the Midwest*, and *Who's Who in America*.

Michael W. Holdren is a senior research scientist in the Atmospheric Science and Applied Technology Department of Battelle in Columbus. His academic background includes a B.S. in chemistry from Washington State University and an M.S. in Environmental Engineering (Air Pollution Studies) from Washington State University. His main interests include determination of human exposure to environmental pollutants, development of sampling and analytical methodologies for measuring air toxics, and investigations of atmospheric chemistry processes — chemical transformation and chemical and physical removal in indoor and outdoor environments.

Mr. Holdren has been actively involved with the U.S. EPA and Department of Defense in developing gas chromatography/mass spectrometric techniques for measuring toxic air pollutants. He has provided technical input and review on many of the U.S. EPA Toxic Organic (TO) documents. He is a co-author of the TO-15 document, which provides guidance for sampling and analyzing the volatile hazardous air pollutants (HAPs) listed in Title III of the 1990 Clean Air Act Amendments (CAAA). He is also working closely with industry in complying with the HAPs emission requirements of the 1990 CAAA. Most of this industrial work focuses on determining air emission factors for chemical processes and consumer products.

He is a member of several professional societies that include the Air and Waste Management Association, Phi Kappa Phi, and Toastmasters International.

Thomas J. Kelly is a senior research scientist in the Atmospheric Science and Applied Technology Department of Battelle in Columbus. He received a B.S. in chemistry from Michigan State University and a Ph.D. in analytical chemistry from the University of Michigan. His research interests include the development and evaluation of analytical methods for environmental pollutants and natural trace species, and the use of monitoring data to determine sources and assess human exposures to airborne pollutants. He has conducted air sampling programs in residential buildings, at industrial facilities, at surface sites, and aboard aircraft; performed data reviews and modeling related to atmospheric chemistry and deposition; and carried out source apportionment of particle- and vapor-phase air pollutants.

Dr. Kelly is the verification testing leader in Battelle's Advanced Monitoring Systems Center, which is part of the U.S. EPA's Environmental Technology Verification (ETV) program. In that capacity he plans, organizes, conducts, and reports on performance evaluations of environmental monitoring technologies. Recent subjects of ETV testing include continuous emission monitors for mercury, and continuous monitors for mass and composition of fine particles in the atmosphere. His other recent activities include development of a real-time monitor for nitrogen dioxide and nitrous acid in indoor air, determination of particle and gas emissions from residential cooking activities, and determination of chemical emission rates from consumer products. He holds two patents on air monitoring devices.

R. Mukund is manager, E-Business and Compliance Systems for General Electric Corporate Environmental Programs. He is currently based in Cincinnati. He received his M.Sc. in Chemistry from the Indian Institute of Technology, Kanpur, and his M.S. and Ph.D. in Environmental Science in Civil Engineering from the University of Illinois, Urbana-Champaign. His current focus is on the development and implementation of Web-based management systems for environmental, health and safety (EHS) processes at global manufacturing and service operations. He is also involved in compliance assurance processes, including facility based self-assessment programs, as well as business and corporate audit programs.

Previously, Dr. Mukund led air compliance management and EHS compliance assurance programs for GE's Power Systems business in Schenectady, NY, and earlier provided Title V air permitting and consulting services for government and industrial clients at ERM-Northeast in Albany, NY.

Prior to his more recent work in environmental consulting and corporate EHS management, Dr. Mukund was part of the Battelle research team that conducted a series of studies relating to hazardous air pollutants, including surveys of ambient air measurement methods, chemical mass balance modeling of the sources of HAPs in urban air, and dispersion modeling studies.

Dr. Mukund is a member of the Air and Waste Management Association.

Table of Contents

Chapter 1

Hazardous Air Pollutants: A Brief Introduction

1.1 Background..1

1.2 The List of Hazardous Air Pollutants 2

1.3 Impact of the HAPs List 8

1.4 Organization of Information in this Book 9

References 10

Chapter 2

The Title III Hazardous Air Pollutants: Classification and Basic Properties

2.1 The 188 Hazardous Air Pollutants: Diversity and Derivation...............11

2.2 Some Common Features of the Title III HAPs.................................11

2.3 Chemical and Physical Properties of the 188 HAPS 12

2.4 Polarizability and Water Solubility as Defining Characteristics of Polar and Nonpolar VOCs 13

Appendix 23

Chapter 3

Measurement Methods for the 188 Hazardous Air Pollutants in Ambient Air

3.1 Introduction..55

3.2 Background...56

3.3 Survey Approach 57

3.4 Status of Current Methods 59

3.5 HAPs Method Development: Future Directions...............................60

3.6 Summary...62

References 62

Appendix 65

Chapter 4

Concentrations of the 188 HAPs in Ambient Air

4.1 Introduction..127

4.2 Survey Procedures 127

4.3 Ambient Air Concentrations of HAPs 129

4.4 Data Gaps 131

4.5 Recent Data for High Priority HAPs...134

4.6 Summary...134

References 134

Appendix 136

Chapter 5
Atmospheric Transformation Products of Clean Air Act Title III Hazardous Air Pollutants

5.1   Introduction...175
5.2   Experimental Approaches for the Study of HAP Transformations...................176
5.3   Hazardous Air Pollutant Transformations...179
5.4   Transformations of 33 high priority HAPs..182
5.5   Transformations of Other Atmospheric Chemicals 183
5.6   Summary...184
References 185
Appendix 187

Index 225

1 Hazardous Air Pollutants: A Brief Introduction

1.1 BACKGROUND

Because of the potential public health implications, the importance of toxic air pollutants in ambient air has been recognized to some degree for many years. Efforts to "regulate" human activities resulting in the production of ambient air pollutants probably date back many centuries, even as the combustion of fossil fuels and air pollution from other organized human activities began having a noticeable impact on the environment. Schemes to classify ambient air pollutants by their human health impacts have evolved with increasing sophistication in this century, culminating in the past 25 years or so with the identification of *toxicity* as a key parameter in identifying ambient air pollutants with serious adverse human health effects, thus warranting their mitigation in some manner.

The definition of a toxic or hazardous air pollutant, however, has had a checkered history as it has evolved in both the scientific community and through the legislative and regulatory process. Patrick[1] provides a detailed discussion of the evolution of the term "toxic air pollutant" through the legislative process in the U.S. He also discusses the formal coining of the term "hazardous air pollutant" in the Clean Air Act of 1970 to represent a group of air pollutants capable of causing adverse health effects, and identified specifically for regulatory oversight in addition to common (or "criteria") pollutants perceived at the time to arise from more ubiquitous sources.

The Clean Air Act Amendments (CAAA) of 1990[2] expanded on the theme of distinguishing among ambient air pollutants by their inherent toxicity, identifying by statute a list of Hazardous Air Pollutants (HAPs). A number of the listed HAPs comprised groups of, rather than individual, chemical substances. The 1990 amendments significantly expanded the focus on toxic air pollutants, and resulted in major regulatory initiatives by the U.S. Environmental Protection Agency (EPA) that have been directed at reducing the emissions of the listed HAPs from stationary sources.

Broadly, the U.S. Congress, through the 1990 amendments,[2] rewrote Section 112 of the Clean Air Act to substantially reduce emissions of HAPs with the intent to provide an "ample margin of safety to protect public health." Key provisions of the revised Section 112 include identification of source categories (industries) that emit one or more of the listed HAPs; schedules for the promulgation of maximum achievable control technology (MACT) or technology-based emission standards for "major" stationary sources by industry category; subsequent risk-based standards that would be triggered if the technology-based standards are not sufficiently protective; and a variety of other directives, including a study of area (non-major stationary or mobile) source HAP emissions, and the development of a strategy to reduce cancer incidence due to HAP emissions of urban area sources by 75%.

The focus on quantifying and mitigating the health risks of the toxic air pollutants that constitute the HAPs list necessitates a depth of knowledge of these pollutants that goes considerably beyond most previous requirements. In particular, the requirements of Section 112(k), "Area Source Program, including National Strategy," require a well-developed database of information on HAP chemical and physical properties as they relate to their presence in ambient air, as well as information on their lifetimes and transformations in ambient air and the availability of measurement methods to quantify current and future concentrations of HAPs in ambient air.

This book represents the consolidation of a series of studies conducted by the authors to support the EPA's mission of understanding and quantifying the health risks from HAPs. The studies were focused on various aspects of the presence of HAPs in ambient air, and addressed chemical and physical properties, currently available measurement methods, the current database of information on the measured ambient concentrations of HAPs in urban areas of the U.S., and our current understanding of the atmospheric transformation products and lifetimes of the HAPs. In contrast to other published handbooks and reference literature on the HAPs, this book is focused on presenting the current state of information on the *presence of the HAPs in ambient air*, as distinct from information on HAPs emission sources, emission measurement methods, control technology and regulatory initiatives and policy. The purpose of this book is to provide readers with a convenient compilation of the information currently available, enabling them to assess the risks posed by HAPs in ambient air, to conduct qualitative comparisons between measured ambient levels of HAPs at specific sites, to guide in understanding the basic chemical and physical properties of the HAPs, and to identify critical research needs at this juncture.

1.2 THE LIST OF HAZARDOUS AIR POLLUTANTS

The term "hazardous air pollutant" was formalized in the 1970 Clean Air Act to mean a pollutant that was not a "criteria" pollutant (as defined further below) and one "which may reasonably be anticipated to result in an increase in mortality or an increase in serious irreversible, or incapacitating reversible, illness." By contrast, "criteria" pollutants were defined as those that "may reasonably be anticipated to endanger public health or welfare," and whose presence in ambient air results from "numerous or diverse mobile or stationary sources." Clearly, "hazardous air pollutant," as defined above, contains an implicit reference to the toxicity of the pollutant, with the further qualification that pollutants such as elemental lead classified as "criteria" pollutants do not qualify as "hazardous air pollutants," regardless of their inherent toxicity. The peculiarities of this definition, which are furthered in other ways in the statutory HAPs list in the 1990 CAAA, must consequently be considered when the terms "hazardous air pollutant" and "toxic air pollutant" are used interchangeably.

Although the 1970 act identified only three initial pollutants of concern — mercury, beryllium, and asbestos — the EPA was directed to list substances that met the definition of a hazardous air pollutant, and to develop regulatory emission standards. Between 1970 and 1990, the EPA listed vinyl chloride, benzene, radionuclides, inorganic arsenic, and coke oven emissions as hazardous air pollutants, with the latter three following the passage of the 1977 CAAA and the directives contained therein on these substances. Although a number of other substances were considered by EPA as potential hazardous air pollutants, no other formal listings were made before the passage of the 1990 Clean Air Act Amendments on November 15, 1990.

EPA's slow pace in listing additional substances as hazardous air pollutants frustrated the U.S. congress, especially considering that EPA-classified carcinogens such as chloroform, formaldehyde, carbon tetrachloride, and polychlorinated biphenyls (PCBs) remained unregulated even 20 years after the 1970 Act authorized the listing and regulation of hazardous air pollutants. Congress, through the 1990 Amendments to the Clean Air Act, therefore created a statutory list in Section 112(b) of 188 HAPs, consisting of both individual chemicals and groups of chemical substances. Table 1.1 provides the current list of the 188 HAPs along with explanatory notes, as of December 10, 2001, from the EPA Air Toxics Website (http://www.epa.gov/ttn/atw/188polls.html).

The derivation of the list of 188 HAPs is described in detail by Patrick.[1] Briefly, it originated from a list of 224 chemicals proposed in Congress by Senator George Mitchell in 1988. That list drew from the more than 300 chemicals from Section 313 of the Emergency Planning and Community Right to Know Act of 1986 (EPCRA); over 100 chemicals from Section 104 of the Comprehensive Emergency Response and Compensation Liability Act (CERCLA or Superfund);

TABLE 1.1
Hazardous Air Pollutants Under Clean Air Act Section 112(b)

Chemical Abstracts Service Number (CAS No.)	Pollutant Name
75-07-0	Acetaldehyde
60-35-5	Acetamide
75-05-8	Acetonitrile
98-86-2	Acetophenone
53-96-3	2-Acetylaminofluorene
107-02-8	Acrolein
79-06-1	Acrylamide
79-10-7	Acrylic acid
107-13-1	Acrylonitrile
107-05-1	Allyl chloride
92-67-1	4-Aminobiphenyl
62-53-3	Aniline
90-04-0	o-Anisidine
1332-21-4	Asbestos
71-43-2	Benzene (including benzene from gasoline)
92-87-5	Benzidine
98-07-7	Benzotrichloride
100-44-7	Benzyl chloride
92-52-4	Biphenyl
117-81-7	Bis(2-ethylhexyl)phthalate (DEHP)
542-88-1	Bis(chloromethyl) ether
75-25-2	Bromoform
106-99-0	1,3-Butadiene
156-62-7	Calcium cyanamide
105-60-2	Caprolactam (removed 6/18/96, 61FR30816)
133-06-2	Captan
63-25-2	Carbaryl
75-15-0	Carbon disulfide
56-23-5	Carbon tetrachloride
463-58-1	Carbonyl sulfide
120-80-9	Catechol
133-90-4	Chloramben
57-74-9	Chlordane
7782-50-5	Chlorine
79-11-8	Chloroacetic acid
532-27-4	2-Chloroacetophenone
108-90-7	Chlorobenzene
510-15-6	Chlorobenzilate
67-66-3	Chloroform
107-30-2	Chloromethyl methyl ether
126-99-8	Chloroprene
1319-77-3	Cresol/Cresylic acid (mixed isomers)
95-48-7	o-Cresol
108-39-4	m-Cresol
106-44-5	p-Cresol
98-82-8	Cumene
N/A	2,4-D (2,4-Dichlorophenoxyacetic Acid) (including salts and esters)

TABLE 1.1 (CONTINUED)
Hazardous Air Pollutants Under Clean Air Act Section 112(b)

Chemical Abstracts Service Number (CAS No.)	Pollutant Name
72-55-9	DDE (1,1-dichloro-2,2-bis(p-chlorophenyl)ethylene)
334-88-3	Diazomethane
132-64-9	Dibenzofuran
96-12-8	1,2-Dibromo-3-chloropropane
84-74-2	Dibutyl phthalate
106-46-7	1,4-Dichlorobenzene
91-94-1	3,3'-Dichlorobenzidine
111-44-4	Dichloroethyl ether (Bis[2-chloroethyl]ether)
542-75-6	1,3-Dichloropropene
62-73-7	Dichlorvos
111-42-2	Diethanolamine
64-67-5	Diethyl sulfate
119-90-4	3,3'-Dimethoxybenzidine
60-11-7	4-Dimethylaminoazobenzene
121-69-7	N,N-Dimethylaniline
119-93-7	3,3'-Dimethylbenzidine
79-44-7	Dimethylcarbamoyl chloride
68-12-2	N,N-Dimethylformamide
57-14-7	1,1-Dimethylhydrazine
131-11-3	Dimethyl phthalate
77-78-1	Dimethyl sulfate
N/A	4,6-Dinitro-o-cresol (including salts)
51-28-5	2,4-Dinitrophenol
121-14-2	2,4-Dinitrotoluene
123-91-1	1,4-Dioxane (1,4-Diethyleneoxide)
122-66-7	1,2-Diphenylhydrazine
106-89-8	Epichlorohydrin (l-Chloro-2,3-epoxypropane)
106-88-7	1,2-Epoxybutane
140-88-5	Ethyl acrylate
100-41-4	Ethylbenzene
51-79-6	Ethyl carbamate (Urethane)
75-00-3	Ethyl chloride (Chloroethane)
106-93-4	Ethylene dibromide (Dibromoethane)
107-06-2	Ethylene dichloride (1,2-Dichloroethane)
107-21-1	Ethylene glycol
151-56-4	Ethyleneimine (Aziridine)
75-21-8	Ethylene oxide
96-45-7	Ethylene thiourea
75-34-3	Ethylidene dichloride (1,1-Dichloroethane)
50-00-0	Formaldehyde
76-44-8	Heptachlor
118-74-1	Hexachlorobenzene
87-68-3	Hexachlorobutadiene
N/A	1,2,3,4,5,6-Hexachlorocyclohexane (all stereo isomers, including lindane)
77-47-4	Hexachlorocyclopentadiene
67-72-1	Hexachloroethane
822-06-0	Hexamethylene diisocyanate

TABLE 1.1 (CONTINUED)
Hazardous Air Pollutants Under Clean Air Act Section 112(b)

Chemical Abstracts Service Number (CAS No.)	Pollutant Name
680-31-9	Hexamethylphosphoramide
110-54-3	Hexane
302-01-2	Hydrazine
7647-01-0	Hydrochloric acid (Hydrogen chloride)
7664-39-3	Hydrogen fluoride (Hydrofluoric acid)
123-31-9	Hydroquinone
78-59-1	Isophorone
108-31-6	Maleic anhydride
67-56-1	Methanol
72-43-5	Methoxychlor
74-83-9	Methyl bromide (Bromomethane)
74-87-3	Methyl chloride (Chloromethane)
71-55-6	Methyl chloroform (1,1,1-Trichloroethane)
78-93-3	Methyl ethyl ketone (2-Butanone)
60-34-4	Methylhydrazine
74-88-4	Methyl iodide (Iodomethane)
108-10-1	Methyl isobutyl ketone (Hexone)
624-83-9	Methyl isocyanate
80-62-6	Methyl methacrylate
1634-04-4	Methyl tert-butyl ether
101-14-4	4,4'-Methylenebis(2-chloroaniline)
75-09-2	Methylene chloride (Dichloromethane)
101-68-8	4,4'-Methylenediphenyl diisocyanate (MDI)
101-77-9	4,4'-Methylenedianiline
91-20-3	Naphthalene
98-95-3	Nitrobenzene
92-93-3	4-Nitrobiphenyl
100-02-7	4-Nitrophenol
79-46-9	2-Nitropropane
684-93-5	N-Nitroso-N-methylurea
62-75-9	N-Nitrosodimethylamine
59-89-2	N-Nitrosomorpholine
56-38-2	Parathion
82-68-8	Pentachloronitrobenzene (Quintobenzene)
87-86-5	Pentachlorophenol
108-95-2	Phenol
106-50-3	p-Phenylenediamine
75-44-5	Phosgene
7803-51-2	Phosphine
7723-14-0	Phosphorus
85-44-9	Phthalic anhydride
1336-36-3	Polychlorinated biphenyls (Aroclors)
1120-71-4	1,3-Propane sultone
57-57-8	β-Propiolactone
123-38-6	Propionaldehyde
114-26-1	Propoxur (Baygon)
78-87-5	Propylene dichloride (1,2-Dichloropropane)
75-56-9	Propylene oxide

TABLE 1.1 (CONTINUED)
Hazardous Air Pollutants Under Clean Air Act Section 112(b)

Chemical Abstracts Service Number (CAS No.)	Pollutant Name
75-55-8	1,2-Propylenimine (2-Methylaziridine)
91-22-5	Quinoline
106-51-4	Quinone (p-Benzoquinone)
100-42-5	Styrene
96-09-3	Styrene oxide
1746-01-6	2,3,7,8-Tetrachlorodibenzo-p-dioxin
79-34-5	1,1,2,2-Tetrachloroethane
127-18-4	Tetrachloroethylene (Perchloroethylene)
7550-45-0	Titanium tetrachloride
108-88-3	Toluene
95-80-7	Toluene-2,4-diamine
584-84-9	2,4-Toluene diisocyanate
95-53-4	o-Toluidine
8001-35-2	Toxaphene (chlorinated camphene)
120-82-1	1,2,4-Trichlorobenzene
79-00-5	1,1,2-Trichloroethane
79-01-6	Trichloroethylene
95-95-4	2,4,5-Trichlorophenol
88-06-2	2,4,6-Trichlorophenol
121-44-8	Triethylamine
1582-09-8	Trifluralin
540-84-1	2,2,4-Trimethylpentane
108-05-4	Vinyl acetate
593-60-2	Vinyl bromide
75-01-4	Vinyl chloride
75-35-4	Vinylidene chloride (1,1-Dichloroethylene)
1330-20-7	Xylenes (mixed isomers)
95-47-6	o-Xylene
108-38-3	m-Xylene
106-42-3	p-Xylene
	Antimony Compounds
	Arsenic Compound (inorganic including arsine)
	Beryllium Compounds
	Cadmium Compounds
	Chromium Compounds
	Cobalt Compounds
	Coke Oven Emissions
	Cyanide Compounds[1]
	Glycol ethers[2]
	Lead Compounds
	Manganese Compounds
	Mercury Compounds
	Fine mineral fibers[3]
	Nickel Compounds
	Polycyclic Organic Matter[4]
	Radionuclides (including radon)[5]
	Selenium Compounds

TABLE 1.1 (CONTINUED)
Hazardous Air Pollutants Under Clean Air Act Section 112(b)

NOTE: For all listings above that contain the word "compounds" and for glycol ethers, the following applies: Unless otherwise specified, these listings are defined as including any unique chemical substance that contains the named chemical (i.e., antimony, arsenic, etc.) as part of that chemical's infrastructure.

[1] X'CN where X = H' or any other group where a formal dissociation may occur, e.g., KCN or Ca(CN).

[2] On January 12,1999 (64FR1780), the EPA proposed to modify the definition of glycol ethers to exclude surfactant alcohol ethoxylates and their derivatives (SAED). On August 2, 2000, (65FR47342), the EPA published the final action.

This action deletes each individual compound in a group called the surfactant alcohol ethoxylates and their derivatives from the glycol ethers category in the list of hazaardous air pollutants (HAP) established by Section 112(b) (1) of the Clean Air Act (CAA). EPA also made conforming changes in the definition of glycol ethers with respect to the designation of hazardous substances under the Comprehensive Environmental Response, Compensation and Liability Act (CERCLA).

"The following definition of the glycol ethers category of hazardous air pollutants applies instead of the definition set forth in 42 U.S.C. 7412(b) (1), footnote 2: Glycol ethers include mono- and di-ethers of ethylene glycol, diethylene glycol, and triethylene glycol R- (OCH2CH2) n-OR'

where:

n = 1, 2 or 3
R = alkyl C7 or less, or phenyl or alkyl substituted phenyl
R' = H, or alkyl C7 or less, or carboxylic acid ester, sulfate, phosphate, nitrate or sulfonate.

[3] Under review.
[4] Under review.
[5] A type of atom that spontaneously undergoes radioactive decay.

"This draft list includes current EPA staff recommendations for technical corrections and clarifications of the hazardous air pollutants (HAP) list in Section 112(b)(1) of the Clean Air Act. This draft has been distributed to apprise interested parties of potential future changes in the HAP list and is informational only. The recommended revisions of the current HAP list which are included in this draft do not themselves change the list as adopted by Congress and have no legal effect. EPA intends to propose specific revisions of the HAP list, including any technical corrections or clarifications of the list, only through notice and comment rulemaking."

and chemicals regulated by various states, such as California, and tracked by the EPA's National Air Toxics Information Clearinghouse (NATICH).

Using the Mitchell list and other data sources, EPA staff recommended the addition of eight and the removal of 51 chemicals.[1] The criteria they used included toxicity, air pollution potential, a previous listing under the Clean Air Act, and whether the chemicals would be regulated under any other provisions of the Act. This resulted in a list containing 181 chemicals.

The list was further refined by EPA on the basis of additional toxicity, emissions, health, and carcinogenicity data, and based on the number of states regulating a chemical through acceptable ambient concentrations.[1] The final list contained 191 HAPs that were included in the bill passed by both houses of Congress in 1991. While undergoing House-Senate conference committee review, however, hydrogen sulfide and ammonia were removed from the list, so that the eventual statutory 112(b) list contained 189 HAPs. The 1990 Amendments bill signed by President George H.W. Bush erroneously included hydrogen sulfide and listed 190 HAPs, even though hydrogen sulfide had been removed in the conference committee; this error was rectified by Congress in late 1991, removing hydrogen sulfide from the list, and an amendment to correct the error was signed by President Bush, resulting in the list of 189 HAPs. Later, EPA was petitioned to delete caprolactam (CAS No. 105-60-2) from the list. After considering all public comments and completing a comprehensive review of available data, EPA published a final rule delisting caprolactam in 1996 (61 FR 30816; June 18, 1996).

There are several noteworthy implications regarding the origin and evolution of the final HAPs list. First, in evaluating the HAPs list as a means of assessing health risks from ambient toxic air pollutants, it must be recognized that the list originated from a very diverse background that included both subjective and political considerations, and it lacks well-defined quantitative listing criteria. Consequently, a number of toxic air pollutants with well-recognized adverse health effects, such as hydrogen sulfide, environmental tobacco smoke, crystalline silica and nitric acid, are not included in the list. At the same time, the list includes a number of substances, such as titanium tetrachloride, phosphorus, and diazomethane, that are unlikely to exist in ambient air because of their extreme reactivity. In essence, the HAPs list is neither a prioritized nor an all-inclusive list of toxic air pollutants. Second, the list actually includes thousands of individual chemical substances, when considering the 17 listed groups of substances, even though not all individual members of a group are necessarily as toxic as others in the group. Finally, the diversity of the group virtually guarantees great variability in the state of existing information on the properties and presence of individual HAP substances in ambient air.

Although the 1990 amendments did not provide a quantitative definition or process by which a HAP could be identified from among the universe of potential toxic air pollutants (just as the 1970 Act did not), Section 112(b)(2) of the Act does include a directive to EPA to periodically (and when petitioned) review and revise the list as necessary by adding pollutants that "… present, or may present, through inhalation or other routes of exposure, a threat of human health effects (including, but not limited to, substances which are known to be, or may reasonably be anticipated to be, carcinogenic, mutagenic, teratogenic, neurotoxic, which cause reproduction dysfunction, or which are acutely or chronically toxic) or adverse environmental effects whether through ambient concentrations, bioaccumulation, deposition, or otherwise…." As noted earlier, EPA issued final notice of the removal of caprolactam from the 112(b) list in June 1996. As the footnotes to Table 1.1 indicate, however, EPA is continuing to review and develop clarifications to the HAPs list on a case-by-case basis.

1.3   IMPACT OF THE HAPS LIST

Despite the tangled evolution of the list of 188 HAPs, that list serves as the starting point in efforts to reduce human health risks from toxic air pollutants. As part of an overall effort to reduce toxic air pollution, EPA has formulated an Integrated Urban Air Toxics Strategy (IUATS),[3] under the authority of Sections 112(k) and 112 (c)(3) of the Clean Air Act. The IUATS is intended to reduce emissions and health risks of air pollution in urban areas, where the variety of sources and density of population can result in substantial human health risks. The strategy addresses emission reductions from 29 categories of area sources; incorporates risk reduction goals such as the 75% reduction in area source cancer incidence noted above; and also describes mobile source emission reductions developed under other sections of the Act. A key product of the IUATS is the designation of 33 of the 188 HAPs as those causing the greatest potential risks to public health in urban areas. That set of 33 high priority HAPs is shown in Table 1.2, and includes 30 HAPs identified as being mostly attributable to area sources and three others less attributable to area sources.

TABLE 1.2
33 High Priority HAPs Designated Under the Integrated Urban Air Toxics Strategy*

Acetaldehyde	Formaldehyde
Acrolein	Hexachlorobenzene
Acrylonitrile	Hydrazine
Arsenic compounds	Lead compounds
Benzene	Manganese compounds
Beryllium compounds	Mercury compounds
1,3-Butadiene	Methylene chloride
Cadmium compounds	Nickel compounds
Carbon tetrachloride*	Polychlorinated biphenyls (PCBs)
Chloroform	Polycyclic organic matter (POM)
Chromium compounds	Propylene dichloride (1,2-dichloropropane)
Coke oven emissions*	Quinoline
1,3-Dichloropropene	1,1,2,2-Tetrachloroethane
2,3,7,8-TCDD (dioxin)	Tetrachloroethylene (perchloroethylene)
Ethylene dibromide* (1,2-dibromoethane)	Trichloroethylene
Ethylene dichloride (1,2-dichloroethane)	Vinyl chloride
Ethylene oxide	

*Among the 33 listed, these three compounds have the lowest proportions of emissions from area sources.

The 33 high priority HAPs listed in Table 1.2 will be the focus of emission reduction efforts, ambient measurements, atmospheric modeling, and health risk assessments under the IUATS. Because of the prominence of these 33 HAPs, particular attention is devoted to them throughout this book.

The list of 188 HAPs is also incorporated into lists of toxic air pollutants established by states and other organizations. For example, the California Air Resources Board has established a list of toxic air contaminants (TACs) that mirrors the list of 188 HAPs.[4] An additional list of 50 candidate TACs has also been established, including species notably absent from the HAPs list, such as hydrogen sulfide, environmental tobacco smoke, crystalline silica, and nitric acid.[4] The candidate TACs are in the process of being evaluated for addition to the TAC list. Similarly, the Texas Natural Resource Conservation Commission has established the effects screening level (ESL) list,[5] which consists of approximately 1,900 toxic chemicals, including the 188 HAPs. That list is used in considering the potential for health or vegetative effects, odor, or property damage. Other states rely on the list of 188 HAPs as their primary set of toxic air contaminants, with specific selections relevant to their own populations. For example, the New Jersey Department of Environmental Protection's Air Toxics Program[6] publishes a list of the 25 air pollutants of greatest concern in that state, based on estimates from EPA's Cumulative Exposure Project.[7] Of those 25 air pollutants, 22 are among the 33 high priority HAPs listed in Table 1.2, confirming their importance as sources of human health risks in urban areas. Thus, although the presence of some species on the HAPs list may be questioned, the 188 HAPs are a *de facto* initial set of toxic air pollutants for many states and agencies.

1.4 ORGANIZATION OF INFORMATION IN THIS BOOK

This book begins (Chapter 2, Classification and Basic Properties) with a grouping of the HAPs into classes of compounds. Following the division of the 188 HAPs into organic and inorganic compounds, vapor pressure (VP, mm Hg at 25° C), boiling point (or melting point), and water solubility (at 25° C) data are presented for all HAPs. Vapor pressure data are used to categorize the 188 HAPs, using quantitative (but subjective) VP criteria to define very volatile, volatile, semivolatile, and nonvolatile compounds.

Chapter 3, Measurement Methods, surveys the current status of ambient measurement methods for the 188 HAPs in Title III of the Clean Air Act Amendments. Measurement methods for the HAPs are identified by reviews of established methods and by literature searches for pertinent research techniques. Methods are segregated by their degree of development into *applicable*, *likely*, and *potential* methods. This survey includes more than 300 methods, applicable to one or more HAPs, in varying stages of method development. The results of the methods survey are tabulated for each of the 188 HAPs, and recommendations for method development initiatives are included.

Chapter 4, Ambient Air Concentrations, presents the results of surveys of ambient air concentrations of the 188 HAPs, illustrating what information is currently available and what is lacking for assessing the public health risks from these HAPs. The ambient concentration survey consists of an extensive table that lists for each HAP the locations, dates, number of samples, means (or medians), and ranges of ambient concentrations, along with citations to the pertinent literature. Comments on the sampling and the number of nondetects reported in ambient samples are also included in the table.

Chapter 5 reviews the methods used to study atmospheric transformations of toxic air pollutants and discusses the concept of atmospheric lifetime of chemicals. It then presents the results of a literature survey on the transformation products, lifetimes and removal pathways for the HAPs, a survey that resulted in 190 literature citations relating to HAP transformations. The chapter concludes with a discussion of gaps in our knowledge of HAP transformations, and recommends that additional information be gathered on the transformation products and removal rates of certain HAPs.

REFERENCES

1. Patrick, D.R., Toxic air pollutants and their sources, in *Toxic Air Pollution Handbook*, Patrick, D.R., Ed., Van Nostrand Reinhold, New York, 33, 1994.
2. The Clean Air Act (as amended through December 31, 1990); 42 U.S. Code, 74.01–76.26; U.S. Government Printing Office, Washington, D.C., April 1991.
3. U.S. EPA, National Air Toxics Program: The Integrated Urban Strategy, 64 FR 38705, July 19, 1999. Available at: http://www.epa.gov/ttnatw01/urban/urbanpg.html.
4. California Air Resources Board, California Air Toxics Program, available at http://www.arb.ca.gov/html/brochure/airtoxic.htm. TAC list available at http://www.arb.ca.gov/toxics/quickref.htm.
5. Texas Natural Resource Conservation Commission, Effects Screening Levels List, available at http://www.tnrcc.state.tx.us/permitting/tox/es/.html. Updated July 19, 2000.
6. New Jersey Department of Environmental Protection, Air Toxics in New Jersey, available at http://www.state.nj.us/dep/airmon/airtoxics. Updated July 30, 2001.
7. U.S. EPA, Cumulative Exposure Project, Air Toxics Component, available at http://www.epa.gov/cumulativeexposure/air/air.htm. Updated April 19, 1999.

2 The Title III Hazardous Air Pollutants: Classification and Basic Properties

2.1 THE 188 HAZARDOUS AIR POLLUTANTS: DIVERSITY AND DERIVATION

The 188 chemicals that constitute the Title III HAPs are a remarkably diverse collection of individual chemicals and generic compound groups, and include industrial chemicals and intermediates, pesticides, chlorinated and hydrocarbon-based solvents, metals, combustion byproducts, chemical groups such as polychlorinated biphenyls (PCBs), and mixed chemicals such as coke-oven emissions. Some of the HAPs, such as volatile organic compounds (VOCs), are common air pollutants, but many others — assigned to the list based on their recognized toxicity in workplace environments — are not typically measured or even considered in ambient air. About one third of the Title III HAPs are semivolatile organic compounds (SVOCs); that is, they may exist in both vapor and particle phases in the atmosphere. Several of the listed HAPs are not single compounds, but rather complex mixtures or groups of chemicals spanning broad ranges of chemical and physical properties. A few, such as titanium tetrachloride, phosphorus, and diazomethane, are unlikely to exist in ambient air because of their reactivity. The survey has information on the basic properties and chemical/volatility group classifications of the 188 Title III HAPs. The survey also includes information on the major sources of the pollutants[1] and identifies the 33 urban air toxics that, under EPA's Integrated Urban Air Toxics Strategy[2] "pose the greatest potential health threat in urban areas."

2.2 SOME COMMON FEATURES OF THE TITLE III HAPS

To facilitate the collection of information about the 188 HAPs, it has been found convenient to arrange them according to their chemical or physical properties. For example, in the ambient concentration and transformation surveys[3–5] (described in Chapters 4 and 5), the 188 HAPs were organized into 10 categories of chemically similar substances. The categories range in size from as few as two to as many as 49 HAPs, and consist of nitrogen- (49), oxygen- (39), and halogen-containing (27) hydrocarbons, as well as inorganics (23), aromatics (18), pesticides (15), halogenated aromatics (8), phthalates (4), hydrocarbons (3) and sulfates (2).

For the survey of the current status of ambient measurement methods[1,2] (described in Chapter 3), key physical and chemical properties were used to group the 188 HAPs into compound classes before evaluating the applicability of individual measurement methods. This contrasts with most previous studies[8,9] which have generally not considered these properties. The approach commonly taken has been to suggest measurement methods for HAPs based on the perceived similarity of one HAP to another. The HAPs' diversity makes this approach suspect. This book includes a detailed review of physical and chemical properties of the HAPs, which were the basis for the identification and selection of appropriate measurement methods.[6,7] The approach is therefore not simply based on apparent similarities in chemical composition and is designed to avoid the short-comings of previous surveys in identifying appropriate measurement methods for the HAPs.

2.3 CHEMICAL AND PHYSICAL PROPERTIES OF THE 188 HAPS

The chemical and physical properties of interest in this survey are those that affect the sampling and measurement of HAPs in the atmosphere.[10–12] To organize the compilation of properties, the 188 HAPs are first divided into organic and inorganic compounds. This initial distinction was based largely on the designation of chemicals in the *CRC Handbook of Chemistry and Physics*[13] and on the known nature of the HAPs. The primary properties thus assembled for all the HAPs were vapor pressure (VP, mm Hg at 25° C), and boiling point or melting point. The vapor pressure data were the primary factor used to categorize and rank the 188 HAPs, because vapor pressure indicates the likely physical state of a chemical in the atmosphere, with boiling point a secondary ranking factor.

Once ranked according to vapor pressure, the HAPs were further grouped using quantitative (but subjective) VP criteria to define very volatile organic and inorganic compounds (i.e., VVOC and VVINC), volatile compounds (VOC and VINC), semivolatile compounds (SVOC and SVINC), and nonvolatile compounds (NVOC and NVINC). The vapor pressure criteria corresponding to each of these HAPs volatility classes are shown in Table 2.1. The vapor pressure criteria shown are the same as those used in previous such categorizations, except for the very volatile categories (VVOC and VVINC). This study denoted as very volatile any compound with a vapor pressure greater than 380 mm Hg (i.e., half an atmosphere); previous categorizations used a somewhat lower criterion of 10 kPa (i.e., 0.099 atm). The vapor pressure criteria are somewhat arbitrary, and compounds with vapor pressures near the boundary values generally fall into "gray areas" that merely define gradual transitions from one volatility class to the next.

Table 2.1 shows that the largest classes are the volatile and semivolatile compounds. Organic compounds (165 chemicals) predominate over inorganic compounds (23 chemicals) in the HAPs list. Inorganic elements and compounds compose the majority of the nonvolatile class of HAPs, i.e., those compounds found exclusively in the particulate phase in the atmosphere.

For the volatile and very volatile HAPs, further chemical and physical properties were compiled, consisting of electronic polarizability, water solubility, aqueous reactivity, and estimated lifetime relative to chemical reaction or deposition in the atmosphere. These properties were included because they determine the effectiveness with which a HAP may be sampled in the atmosphere,[10,12] and the extent to which atmospheric processes may obscure emissions of HAPs to the atmosphere. Table 2.1 also summarizes the properties reviewed for the various volatility classes of HAPs and the number of HAPs in each class.

Nineteen of the HAPs are listed simply as compound groups (e.g., PCBs). Based on which compounds in each of these groups are most likely to be present in ambient air, these HAPs were classified in multiple volatility classes. For the purposes of the above count of HAPs in each volatility class, compound group HAPs were categorized on the basis of the most volatile species in each group likely to be present in ambient air.

The primary information sources used for the HAPs properties survey are handbooks and databases of chemical and physical properties.[8,13–21] Whenever possible, inconsistencies and errors were corrected by comparisons of data from various sources, and by consultation with EPA staff.

The chemical and physical property data compiled in this work are presented in detail in Table 2.2 (see Appendix following Chapter 2) for the full list of 188 HAPs. Table 2.2 lists the 188 HAPs in the order in which they appear in the Clean Air Act Amendments, along with the Chemical Abstracts Service (CAS) number and properties compiled for each HAP. The successive columns in the table list the HAP name, chemical formula or structure, CAS number, molecular weight (MW), major sources, chemical class, volatility class, vapor pressure (VP), boiling point (BP), and water solubility. A column for comments is also included. The second tabular form is Table 2.3, which presents additional properties for VVOCs and VOCs only. This table includes some of the data from Table 2.2,

TABLE 2.1
Summary of HAP Categories with Defined Vapor Pressure Ranges, Relevant Properties, and Number of HAPs in Each Category

Volatility Class[*]	Vapor Pressure Range (mm Hg at 25° C)	No. of HAPs	Relevant Properties
VVOC	>380	15	Vapor pressure; boiling point; water solubility; polarizability; aqueous reactivity; atmospheric lifetime
VVINC	>380	6	Vapor pressure; boiling point; water solubility; polarizability; aqueous reactivity; atmospheric lifetime
VOC	0.1 to 380	82	Vapor pressure; boiling point; water solubility; polarizability; aqueous reactivity; atmospheric lifetime
VINC	0.1 to 380	3	Vapor pressure; boiling point; water solubility; polarizability; aqueous reactivity; atmospheric lifetime
SVOC	10^{-7} to 0.1	63	Vapor pressure; boiling point
SVINC	10^{-7} to 0.1	2	Vapor pressure; boiling point
NVOC	$<10^{-7}$	5	Vapor pressure; boiling point
NVINC	$<10^{-7}$	12	Vapor pressure; boiling point

[*] VVOC   =   very volatile organic compounds
VVINC   =   very volatile inorganic compounds
VOC   =   volatile organic compounds
VINC   =   volatile inorganic compounds
SVOC   =   semivolatile organic compounds
SVINC   =   semivolatile inorganic compounds
NVOC   =   nonvolatile organic compounds
NVINC   =   nonvolatile inorganic compounds.

but also includes information on electronic polarizability and aqueous reactivity. Table 2.4 provides a separate list identifying the organic and inorganic HAPs in each of the volatility ranges.

2.4   POLARIZABILITY AND WATER SOLUBILITY AS DEFINING CHARACTERISTICS OF POLAR AND NONPOLAR VOCS

Volatile organic compounds in air consist largely of hydrocarbons and oxygenated hydrocarbons, as well as some nitrogen- and sulfur-containing compounds. The oxygenated hydrocarbons, in turn, consist of several compound classes, including alcohols, aldehydes, ketones, ethers, carboxylic acids, etc. For analytical purposes, airborne organic compounds can be considered as either nonpolar (i.e., hydrocarbons) or polar (i.e., compounds containing oxygen, sulfur, nitrogen, etc.).

Nonpolar VOCs can be characterized at the part-per-billion by volume (ppbv) level using currently available methods. However, polar VOCs tend to be difficult to sample and analyze at trace levels because of their chemical reactivity, affinity for metal and other surfaces, and solubility in water.[12] Because polar VOCs include compound classes typically associated with higher polarizabilities, the general classification of the VOCs on the HAPs list was investigated as a function of electronic polarizability (molar refractivity). Polarizabilities were calculated from the relationship:

$$Molar\ Reflectivity = \frac{MW}{\rho} \frac{n^2 - 1}{n^2 + 2}$$

TABLE 2.3
Physical and Chemical Properties of Volatile Organic Compounds in the HAPs List

Compound	CAS No.	Sub-category[1]	VP[2] (mm Hg at 25°C)	Polarizability[3] (cm³/mole)	Water Solubility[2] (g/L at 25°C)	Aqueous Reactivity[4]	Other[5]
Acetaldehyde	75-07-0	VVOC	904	11.6	1000	—	Polar
Acetonitrile	75-05-8	VOC	88.5	11.0	1000	aab	Polar
Acetophenone	98-86-2	VOC	0.44	36.3	6.1	—	Polar
Acrolein	107-02-8	VOC	275	16.2	213	aab	Polar
Acrylamide	79-06-1	VOC	0.007	—	Reacts	—	Polar
Acrylic acid	79-10-7	VOC	4.0	17.4	1000	aab	Polar
Acrylonitrile	107-13-1	VOC	109	15.6	74.5	aab	Polar
Allyl chloride	107-05-1	VOC	369	20.5	3.4	ah	Nonpolar
Aniline	62-53-3	VOC	0.64	30.6	36.0	—	Polar
Benzene	71-43-2	VOC	95.5	26.2	1.8	aab	Nonpolar
Benzyl chloride	100-44-7	VOC	1.2	36.0	Reacts	ah	Nonpolar
Bis(chloromethyl) ether	542-88-1	VOC	30.0 at 22°C	22.8	Reacts	ah	Polar
Bromoform	75-25-2	VOC	5.4	29.8	3.1	aab	Nonpolar
1,3-Butadiene	106-99-0	VVOC	2113	22.4	0.5	aab	Nonpolar
Carbon disulfide	75-15-0	VOC	361	21.5	1.2	—	Nonpolar
Carbon tetrachloride	56-23-5	VOC	114	26.5	0.8	aab	Nonpolar
Carbonyl sulfide	463-58-1	VVOC	9623	12.6	1.2	—	Polar
Catechol	120-80-9	VOC	0.03 at 20°C	32.9	461	aab	Polar
Chloroacetic acid	79-11-8	VOC	0.07	17.6	100 at 20°C	aab	Polar
Chlorobenzene	108-90-7	VOC	12.0	31.1	0.5	aab	Nonpolar
Chloroform	67-66-3	VOC	197	21.4	8.0	aab	Nonpolar
Chloromethyl methyl ether	107-30-2	VOC	187	18.2	Reacts	ah	Polar
Chloroprene	126-99-8	VOC	216	25.2	0.9	aab	Nonpolar
Cresylic acid (cresol isomers)	1319-77-3	VOC	0.1	32.5	19.3	aab	Polar
o-Cresol	95-48-7	VOC	0.2	32.2	25.9	aab	Polar
Cumene	98-82-8	VOC	3.5	40.5	0.05	aab	Nonpolar
Diazomethane	334-88-3	VVOC	2800	—	Reacts	—	Polar
1,2-Dibromo-3-chloropropane	96-12-8	VOC	0.6 at 20°C	36.3	1.2 at 20°C	aab	Nonpolar
1,4-Dichlorobenzene	106-46-7	VOC	1.8	36.3	0.08	aab	Nonpolar
Dichloroethyl ether (Bis(2-chloroethyl)ether)	111-44-4	VOC	1.6	32.0	Reacts	aab	Polar
1,3-Dichloropropene	542-75-6	VOC	34.0	25.5	2.8 at 20°C	ah	Nonpolar
Diethyl sulfate	64-67-5	VOC	0.21	31.6	Reacts	ah	Polar
N,N-Dimethylaniline	121-69-7	VOC	0.8	40.8	1.5	aab	Polar
Dimethylcarbamoyl chloride	79-44-7	VOC	2.0	—	Reacts	ah	Polar
N,N-Dimethylformamide	68-12-2	VOC	2.6	19.9	1000	—	Polar
1,1-Dimethylhydrazine	57-14-7	VOC	157	18.7	Reacts	aab	Nonpolar
Dimethyl sulfate	77-78-1	VOC	0.7	—	28.0 at 18°C	ah	Polar
1,4-Dioxane	123-91-1	VOC	27.0	21.4	1000	aab	Polar
Epichlorohydrin	106-89-8	VOC	12.5	20.5	65.9	aab	Polar
1,2-Epoxybutane	106-88-7	VOC	180	20.3	95.0	ah	Polar
Ethyl acrylate	140-88-5	VOC	38.4	26.6	15.0	aab	Polar
Ethylbenzene	100-41-4	VOC	12.7	35.7	0.2	aab	Nonpolar

TABLE 2.3 (CONTINUED)
Physical and Chemical Properties of Volatile Organic Compounds in the HAPs List

Compound	CAS No.	Sub-category[1]	VP[2] (mm Hg at 25°C)	Polarizability[3] (cm³/mole)	Water Solubility[2] (g/L at 25°C)	Aqueous Reactivity[4]	Other[5]
Ethyl carbamate	51-79-6	VOC	0.3	22.6	480 at 15°C	aab	Polar
Ethyl chloride	75-00-3	VVOC	1202	16.2	5.7 at 20°C	aab	Nonpolar
Ethylene dibromide	106-93-4	VOC	14.3	27.0	4.2	aab	Nonpolar
Ethylene dichloride	107-06-2	VOC	78.7	21.0	8.6	aab	Nonpolar
Ethyleneimine	151-56-4	VOC	213	—	1000	aab	Polar
Ethylene oxide	75-21-8	VVOC	1311	11.2	1000	ah	Polar
Ethylidene dichloride	75-34-3	VOC	227	21.1	5.1	aab	Nonpolar
Formaldehyde	50-00-0	VVOC	3821	8.4	550	aab	Polar
Hexachlorobutadiene	87-68-3	VOC	0.2	49.8	0.003	aab	Nonpolar
Hexachloroethane	67-72-1	VOC	0.6	—	0.05	aab	Nonpolar
Hexane	110-54-3	VOC	151	29.9	0.01	—	Nonpolar
Isophorone	78-59-1	VOC	0.4	42.1	12.0	aab	Polar
Methanol	67-56-1	VOC	118	8.2	1000	aab	Polar
Methyl bromide	74-83-9	VVOC	1646	15.0	15.2	aab	Nonpolar
Methyl chloride	74-87-3	VVOC	4318	11.5	5.3	ah	Nonpolar
Methyl chloroform	71-55-6	VOC	133	26.2	1.5	aab	Nonpolar
Methyl ethyl ketone	78-93-3	VOC	90.0	20.7	256 at 20°C	aab	Polar
Methylhydrazine	60-34-4	VOC	49.6	13.7	Reacts	aab	Nonpolar
Methyl iodide	74-88-4	VVOC	402	19.3	1.4	aab	Nonpolar
Methyl isobutyl ketone	108-10-1	VOC	19.7	30.0	19.0	aab	Polar
Methyl isocyanate	624-83-9	VOC	348 at 20°C	14.0	Reacts	ah	Polar
Methyl methacrylate	80-62-6	VOC	36.1	26.5	15.0	aab	Polar
Methyl tert-butyl ether	1634-04-4	VOC	250	26.2	51.0	aab	Polar
Methylene chloride	75-09-2	VOC	435	16.4	13.2	aab	Nonpolar
Nitrobenzene	98-95-3	VOC	0.25	32.7	1.9	aab	Polar
2-Nitropropane	79-46-9	VOC	17.0	21.6	17.0	aab	Polar
N-Nitroso-N-methylurea	684-93-5	VOC	0.03	—	14.4 at 23° C	ah	Polar
N-Nitrosodimethylamine	62-75-9	VOC	2.7 at 20°C	19.3	1000 at 24° C	aab	Polar
N-Nitrosomorpholine	59-89-2	VOC	0.04 at 20°C	—	1000 at 24° C	aab	Polar
Phenol	108-95-2	VOC	0.35	28.0	82.8	aab	Polar
Phosgene	75-44-5	VVOC	1406	—	Slightly soluble	aab	Polar
1,3-Propane sultone	1120-71-4	VOC	0.27	—	171	ah	Polar
β-Propiolactone	57-57-8	VOC	3.4	15.7	10-50 at 19° C	ah	Polar
Propionaldehyde	123-38-6	VOC	317	16.1	306	aab	Polar
Propylene dichloride	78-87-5	VOC	50.4	25.7	2.7	aab	Nonpolar
Propylene oxide	75-56-9	VVOC	530	15.7	590	ah	Polar
1,2-Propyleneimine	75-55-8	VOC	112 at 20° C	17.6	1000	aab	Polar
Styrene	100-42-5	VOC	6.1	36.4	0.32	aab	Nonpolar
Styrene oxide	96-09-3	VOC	0.3 at 20°C	35.5	3.0 at 20° C	ah	Polar
1,1,2,2-Tetrachloroethane	79-34-5	VOC	4.0	30.7	3.0	ah	Nonpolar
Tetrachloroethylene	127-18-4	VOC	18.6	30.3	0.2	aab	Nonpolar
Toluene	108-88-3	VOC	28.6	31.0	0.53	aab	Nonpolar
1,2,4-Trichlorobenzene	120-82-1	VOC	0.42	41.0	0.05	aab	Nonpolar
1,1,2-Trichloroethane	79-00-5	VOC	22.0	25.9	4.4	aab	Nonpolar
Trichloroethylene	79-01-6	VOC	69.0	25.4	1.1	aab	Nonpolar

TABLE 2.3 (CONTINUED)
Physical and Chemical Properties of Volatile Organic Compounds in the HAPs List

Compound	CAS No.	Sub-category[1]	VP[2] (mm Hg at 25°C)	Polarizability[3] (cm³/mole)	Water Solubility[2] (g/L at 25°C)	Aqueous Reactivity[4]	Other[5]
Triethylamine	121-44-8	VOC	57.9	33.8	74.0	—	Polar
2,2,4-Trimethylpentane	540-84-1	VOC	48.7	39.2	0.002	—	Nonpolar
Vinyl acetate	108-05-4	VOC	115	22.2	20.0 at 20° C	aab	Polar
Vinyl bromide	593-60-2	VVOC	1059	18.9	5.7	aab	Nonpolar
Vinyl chloride	75-01-4	VVOC	2937	15.5	1.1	aab	Nonpolar
Vinylidene chloride	75-35-4	VVOC	600	20.4	2.3	aab	Nonpolar
Xylenes (isomer mixture)	1330-20-7	VOC	8.0	36.1	0.2	aab	Nonpolar
o-Xylene	95-47-6	VOC	6.6	35.8	0.2	aab	Nonpolar
m-Xylene	108-38-3	VOC	8.3	36.0	0.2	aab	Nonpolar
p-Xylene	106-42-3	VOC	8.9	36.0	0.2	aab	Nonpolar

1. VVOC = Very Volatile Organic Compounds (vapor pressure at 25° C >380 mm Hg)

   VOC = Volatile Organic Compounds (0.1< vapor pressure at 25° C <380 mm Hg).

2. Vapor pressure (VP) and water solubility data from: (a) Ref. 16; (b) Ref. 17; (c) R.C. Weast, Ed., *CRC Handbook of Chemistry and Physics*, 59th ed., CRC, Boca Raton, 1979; (d) Ref. 14; (e) Ref. 18; (f) Ref. 21; (g) Ref. 19; (h) Ref. 20.

3. Electronic polarizability = $(MW/\rho)[n^2 - 1]/[n^2 + 2]$ from: E.B. Sansone et al., Prediction of removal of vapors from air by adsorption on activated carbon, *Environ. Sci. Technol.*, 13, 1511-1513 (1979). Values for molecular weight (MW), density (ρ), and refractive index (n) are taken from: (a) R.C. Weast, Ed., *CRC Handbook of Chemistry and Physics*, 59th ed., CRC, Boca Raton, 1979; (b) Ref. 8.

4. Reactivity data from Ref. 15.

   aab = aqueous aerobic biodegradation; ah = aqueous hydrolysis; ab = aerobic biodegradation; h = hydrolysis.

5. Customary classification of VOCs as either Nonpolar or Polar.

where MW = molecular weight; ρ = density; and n = refractive index. Figure 2.1 shows the data generated in this way for the VOCs. This plot ranks the VOCs that are customarily identified as either nonpolar (N) or polar (P) compounds as a function of their electronic polarizability. Figure 2.1 shows that the N and P compounds are well mixed in the ranking on the basis of polarizability. It is evident from this plot that, based on polarizability, there is no clear distinction between the N and P compounds, because both groups of compounds are distributed over the entire polarizability range.

Because of the collection and analysis problems known to arise as a result of the water solubility of certain VOCs, the VOCs were also ranked on the basis of their solubility in water at 25° C. The most useful literature compilations found were those of Keith and Walker,[8] Mackay et al.,[14] and The Physical Properties Database (PHYSPROP) from Syracuse Research Corporation.[18]

Figure 2.2 shows a plot ranking the VOCs as a function of their water solubility. Here, one can see that compounds that have conventionally been identified as nonpolar VOCs are characterized by relatively low water solubilities, whereas compounds that are generally regarded as polar VOCs are characterized by relatively high water solubilities. Classifying VOCs on the basis of their solubility in water therefore provides a more meaningful distinction between polar and nonpolar compounds than does classification on the basis of polarizability.

TABLE 2.4
HAPs Grouped by Volatility Class

VVOCs ($VP_{25°C}$ > 380 mm Hg)	VOCs (0.1 mm Hg < $VP_{25°C}$ < 380 mm Hg)	
Acetaldehyde	Acetonitrile	Isophorone
1,3-Butadiene	Acetophenone	Methanol
Carbonyl sulfide	Acrolein	Methyl chloroform
Diazomethane	Acrylamide	Methyl ethyl ketone
Ethyl chloride	Acrylic acid	Methylhydrazine
Ethylene oxide	Acrylonitrile	Methyl isobutyl ketone
Formaldehyde	Allyl chloride	Methyl isocyanate
Methyl bromide	Aniline	Methyl methacrylate
Methyl chloride	Benzene	Methyl tert-butyl ether
Methyl iodide	Benzyl chloride	Methylene chloride
Phosgene	Bis (chloromethyl) ether	Nitrobenzene
Propylene oxide	Bromoform	2-Nitropropane
Vinyl bromide	Carbon disulfide	N-Nitroso-N-methylurea
Vinyl chloride	Carbon tetrachloride	N-Nitrosodimethylamine
Vinylidene chloride	Catechol	N-Nitrosomorpholine
(Total of 15 HAPs)	Chloroacetic acid	Phenol
	Chlorobenzene	1,3-Propane sultone
	Chloroform	β-Propiolactone
	Chloromethyl methyl ether	Propionaldehyde
	Chloroprene	Propylene dichloride
	Cresol/Cresylic acid (mixed isomers)	1,2-Propylenimine
	o-Cresol	Styrene
	Cumene	Styrene oxide
	1,2-Dibromo-3-chloropropane	1,1,2,2-Tetrachloroethane
	1,4-Dichlorobenzene	Tetrachloroethylene
	Dichloroethyl ether	Toluene
	(Bis[2chloroethyl]ether)	1,2,4-Trichlorobenzene
	1,3-Dichloropropene	1,1,2-Trichloroethane
	Diethyl sulfate	Trichloroethylene
	N,N-Dimethylaniline	Triethylamine
	Dimethylcarbamoyl chloride	2,2,4-Trimethylpentane
	N,N-Dimethylformamide	Vinyl acetate
	1,1-Dimethylhydrazine	Xylene (mixed isomers)
	Dimethyl sulfate	o-Xylene
	1,4-Dioxane	m-Xylene
	Epichlorohydrin	p-Xylene
	1,2-Epoxybutane	*(Total of 82 HAPs)*
	Ethyl acrylate	
	Ethylbenzene	
	Ethyl carbamate	
	Ethylene dibromide	
	Ethylene dichloride	
	Ethyleneimine	
	Ethylidene dichloride	
	Hexachlorobutadiene	
	Hexachloroethane	
	Hexane	

TABLE 2.4 (CONTINUED)
HAPs Grouped by Volatility Class

SVOCs
$(10^{-7}$ mm Hg $< VP_{25°C} < 0.1$ mm Hg)

Acetamide
4-Aminobiphenyl
o-Anisidine
Benzidine

Benzotrichloride
Biphenyl
Bis (2-ethylhexyl)phthalate
Captan
Carbaryl
Chloramben
Chlordane
2-Chloroacetophenone
Chlorobenzilate

m-Cresol
p-Cresol
2,4-D (2,4-Dichloro phenoxyacetic
 acid) (incl. salts and esters)
DDE
Dibenzofurans

Dibutyl phthalate
3,3'-Dichlorobenzidine
Dichlorvos
Diethanolamine
3,3'-Dimethylbenzidine
Dimethyl phthalate
4,6-Dinitro-o-cresol (incl. salts)
2,4-Dinitrophenol
2,4-Dinitrotoluene
1,2-Diphenylhydrazine
Ethylene glycol
Ethylene thiourea
Heptachlor
Hexachlorobenzene
1,2,3,4,5,6-Hexachloro cyclohexane
 (all stereo isomers, incl. Lindane)
Hexachlorocyclo pentadiene
Hexamethylene diisocyanate
Hexamethylphosphoramide
Hydroquinone
Maleic anhydride
Methoxychlor
4,4'-Methylenediphenyl diisocyanate
Naphthalene

4-Nitrophenol
Parathion
Pentachloronitrobenzene
Pentachlorophenol

p-Phenylenediamine
Phthalic anhydride
Polychlorinated biphenyls
Propoxur (Baygon)
Quinoline
Quinone
2,3,7,8-Tetrachlorodibenzo-p-dioxin
Toluene-2,4-diamine
2,4-Toluene diisocyanate

o-Toluidine
Toxaphene (chlorinated camphene)
2,4,5-Trichlorophenol

2,4,6-Trichlorophenol
Trifluralin

Coke oven emissions
Glycol ethers
Polycyclic organic matter
(Total of 63 HAPs)

NVOCs
$(VP_{25°C} < 10^{-7}$ mm Hg)

2-Acetylaminofluorene
3,3'-Dimethoxybenzidine
4-Dimethylaminoazobenzene
4,4'-Methylenebis-(2-chloroaniline)
4,4'-Methylenedianiline
(Total of 5 HAPs)

VVINCs
$(VP_{25°C} > 380$ mm Hg)

Chlorine
Hydrogen fluoride (hydrofluoric acid)
Phosphine
Arsenic compounds (inorganic incl.
 arsine)
Cyanide compounds
Radionuclides (incl. radon)
(Total of 6 HAPs)

VINCs
$(0.1$ mm Hg $< VP_{25°C} < 380$ mm Hg)

Hydrazine
Hydrochloric acid (hydrogen
 chloride)
Titanium tetrachloride
(Total of 3 HAPs)

SVINCs
$(10^{-7}$ mm Hg $< VP_{25°C} < 0.1$ mm Hg)

Phosphorus
Mercury Compounds
(Total of 2 HAPs)

TABLE 2.4 (CONTINUED)
HAPs Grouped by Volatility Class

Note: A number of HAPs can be categorized in more than one volatility class, e.g., mercury compounds in vapor and particulate forms (SVINC and NVINC). In such cases, the HAPs have been assigned in this table based on the vapor pressure of the most volatile species present in ambient air. Thus, for example, mercury compounds have been assigned to the SVINC category using this rationale, although they are present in ambient air in both SVINC and NVINC forms.

NVINCs
$(VP_{25°C} < 10^{-7} \text{ mm Hg})$

Asbestos
Calcium cyanamide
Antimony compounds
Beryllium compounds
Cadmium compounds
Chromium compounds
Cobalt compounds
Lead compounds
Manganese compounds
Fine mineral fibers
Nickel compounds
Selenium compounds
(Total of 12 HAPs)

Note: A number of HAPs can be categorized in more than one volatility class, e.g., mercury compounds in vapor and particulate forms (SVINC and NVINC). In such cases, the HAPs have been assigned in this table based on the vapor pressure of the most volatile species present in ambient air. Thus, for example, mercury compounds have been assigned to the SVINC category using this rationale, although they are present in ambient air in both SVINC and NVINC forms.

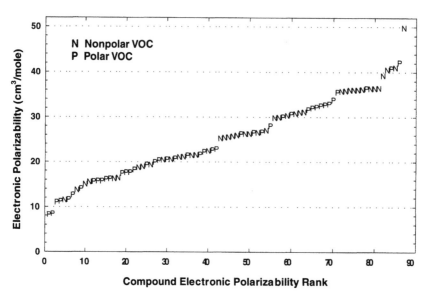

FIGURE 2.1 Ranking of Title III nonpolar (N) and polar (P) VOCs on the basis of electronic polarizability.

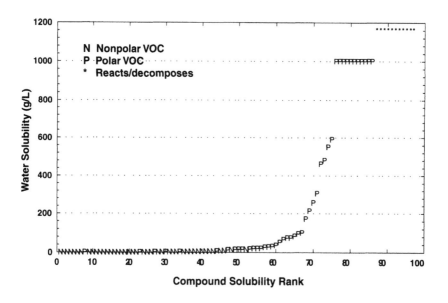

FIGURE 2.2 Ranking of Title III nonpolar (N) and polar (P) VOCs on the basis of water solubility.

REFERENCES

1. Patrick, D.R., Toxic air pollutants and their sources, in *Toxic Air Pollution Handbook*, Patrick, D.R., Ed., Van Nostrand Reinhold, New York, 1994, pp. 33-49.
2. U.S. EPA. National Air Toxics Program: The Integrated Urban Strategy. 64 FR 38705. July 19, 1999. Available at: http://www.epa.gov/ttnatw0l/urban/urbanpg.html.
3. Kelly, T.J. et al., Ambient Concentration Summaries for Clean Air Act Title III Hazardous Air Pollutants, Report EPA-600/R-94-090, U.S. EPA, Research Triangle Park, NC, July 1993.
4. Spicer, C.W. et al., A Literature Review of Atmospheric Transformation Products of Clean Air Act Title III Hazardous Air Pollutants, Report EPA-600/R-94-088, U. S. Environmental Protection Agency, Research Triangle Park, NC, July 1993.
5. Kelly, T.J. et al., Concentrations and transformations of hazardous air pollutants, *Environ. Sci. Technol.*, 28, 378A-387A, 1994.
6. Kelly, T.J. et al., Ambient Measurement Methods and Properties of the 189 Clean Air Act Hazardous Air Pollutants, Report EPA-600/R-94-098, U.S. EPA, Research Triangle Park, NC, March 1994.
7. Mukund, R. et al., Status of ambient measurement methods for hazardous air pollutants, *Environ. Sci. Technol.*, 29, 183A-187A, 1995.
8. Keith, L.H. and Walker, M.M., *EPA's Clean Air Act Air Toxics Database, Volume 1: Sampling and Analysis Methods Summaries*, Lewis, Boca Raton, FL, 1992.
9. Winberry, Jr., W.T., Sampling and analysis under Title III, *Environ. Lab.*, 46-68, June/July 1993.
10. Clements, J.B. and Lewis, R.G., Sampling for Organic Compounds, in *Principles of Environmental Sampling*, Keith, L.H., Ed., American Chemical Society, Washington, D.C., 1987, pp. 287-296.
11. Coutant, R.W. and McClenny, W.A., Competitive adsorption effects and the stability of VOC and PVOC in canisters, in *Proc. 1991 EPA/A&WMA Symp. Measurement of Toxic and Related Air Pollutants*, Publication No. VIP-21, Air and Waste Management Association, Pittsburgh, PA, 1991, pp. 382-388.
12. Lewis, R.G. and Gordon, S.M., Sampling for organic chemicals in air, in *Principles of Environmental Sampling*, Keith, L.H., Ed., 2nd ed., American Chemical Society, Washington, DC, 1996, chap. 23.
13. Lide, D.R., Ed., *CRC Handbook of Chemistry and Physics*, 82nd ed., CRC, Boca Raton, FL, 2001.
14. Mackay, D., Shiu, W.Y. and Ma, K.C., *Illustrated Handbook of Physical–Chemical Properties and Environmental Fate for Organic Chemicals, Volume III: Volatile Organic Chemicals*, Lewis, Chelsea, MI, 1993.
15. Howard, P.H. et al., *Handbook of Environmental Degradation Rates*, Lewis, Chelsea, MI, 1991.
16. Jones, D.L. and Bursey, J., Simultaneous Control of PM-10 and Hazardous Air Pollutants, II: Rationale for Selection of Hazardous Air Pollutants as Potential Particulate Matter, Report EPA-452/R-93/013, U.S. EPA, Research Triangle Park, NC, October 1992.
17. Weber, R.C., Parker, P.A. and Bowser, M., Vapor Pressure Distribution of Selected Organic Chemicals, Report EPA-600/2-81/021, U.S. EPA, Washington, D.C., February 1981.
18. The Physical Properties Database (PHYSPROP), Syracuse Research Corporation, North Syracuse, NY 13212. Available at: http://esc.syrres.com/interkow/PhysProp.htm.
19. The National Institute of Standards and Technology (NIST) Chemistry WebBook. Available at: http://webbook.nist.gov/chemistry/.
20. 1997 Toxic Air Contaminant Identification List – Summaries. California Air Resources Board, California Environmental Protection Agency, Sacramento, CA. Available at: http://www.arb.ca.gov/toxics/tac/toctbl.htm.
21. Available at: http://chemfinder.cambridgesoft.com.

APPENDIX

TABLE 2.2
Properties, Sources/Uses, and Chemical/Volatility Group Classifications of CAAA Title III HAPs (Chemicals shown in italics are high priority urban HAPs)

Compound	Chemical Formula/Structure	CAS No.	MW	Sources[1]	Chemical Class[2]	Volatility Class[3]	VP[4] (mm Hg at 25°C)	BP[4] (°C)	Water Solubility[4] (g/L at 25°C)	Comment
Acetaldehyde; C$_2$H$_4$O	*CH$_3$CHO*	*75-07-0*	*44.0*	*Chemical ind. photochemical reaction prod.*	*Oxy Org*	*VVOC*	*904*	*21*	*1000*	
Acetamide; C$_2$H$_5$NO	CH$_3$C(O)NH$_2$	60-35-5	59.0	Chemical ind.	Nitro Org	SVOC	0.05	222	2250	
Acetonitrile; C$_2$H$_3$N	CH$_3$CN	75-05-8	41.0	Chemical ind.	Nitro Org	VOC	88.5	82	1000	
Acetophenone; C$_8$H$_8$O	O=C—CH$_3$ (phenyl)	98-86-2	120.0	Chemical ind.	Oxy Org	VOC	0.44	202	6.1	
2-Acetylaminofluorene; C$_{15}$H$_{13}$NO	(structure)	53-96-3	223.3	Chemical ind.	Nitro Org	NVOC	9.4×10^{-8}	444	5.3×10^{-3}	
Acrolein; C$_3$H$_4$O	*H$_2$C=CHCHO*	*107-02-8*	*56.0*	*Chemical ind.; photochemical reaction prod.*	*Oxy Org*	*VOC*	*275*	*53*	*213*	*Reactive [5]*
Acrylamide; C$_3$H$_5$NO	CH$_2$CHC(O)NH$_2$	79-06-1	71.0	Chemical ind.	Nitro Org	VOC	7×10^{-3}	125 at 25 mm	2160	Reactive[5]
Acrylic acid; C$_3$H$_4$O$_2$	H$_2$C=CHCO$_2$H	79-10-7	72.0	Chemical ind.	Oxy Org	VOC	4.0	141	1000	

TABLE 2.2 (CONTINUED)

Properties, Sources/Uses, and Chemical/Volatility Group Classifications of CAAA Title III HAPs (Chemicals shown in italics are high priority urban HAPs)

Compound	Chemical Formula/Structure	CAS No.	MW	Sources[1]	Chemical Class[2]	Volatility Class[3]	VP[4] (mmHG at 25° C)	BP[4]	Water Solubility[4]	Comment
Acrylonitrile; C$_3$H$_3$N	*CH$_2$CHCN*	*107-13-1*	*53.0*	*Chemical ind.; plastics*	*Nitro Org*	*VOC*	*109*	*77*	*74.5*	
Allyl chloride; C$_3$H$_5$Cl	CH$_2$=CHCH$_2$Cl	107-05-1	76.5	Chemical ind.	Hal Hydro	VOC	369	45	3.4	
4-Aminobiphenyl; C$_{12}$H$_{11}$N	[structure: biphenyl–NH$_2$]	92-67-1	169.0	Chemical ind.	Nitro Org	SVOC	1.2×10^{-4}	302	0.3	
Aniline; C$_6$H$_7$N	[structure: NH$_2$ on benzene ring]	62-53-3	93.0	Chemical ind.	Nitro Org	VOC	0.64	184	36	
o-Anisidine; C$_7$H$_9$NO	[structure: NH$_2$ and CH$_3$ on benzene ring]	90-04-0	123.0	Chemical ind.	Nitro Org	SVOC	0.07	224	<0.1 at 19°C	Reactive[6]
Asbestos	Silicate minerals of the serpentine and amphibole groups	1332-21-4	--	Insulation	Inorg	NVINC	Very low	Decomposes at 1112°C	Insoluble	

Name	CAS	MW	Source/Use	Class	VOC/SVOC	Vapor pressure	B.P.	Solubility	Reactivity
Benzene; C$_6$H$_6$	71-43-2	78.0	*Chemical ind.; gasoline; smoking*	*Arom*	*VOC*	*95.5*	*80*	*1.8*	
Benzidine; C$_{12}$H$_{12}$N$_2$	92-87-5	184.2	Chemical ind.	Nitro Org	SVOC	7.5 × 10^{-8} at 20°C	402	0.52	
Benzotrichloride; C$_7$H$_5$Cl$_3$	98-07-7	195.5	Chemical ind.	Hal Arom	SVOC	0.41	213	Reacts	Reactive[5]
Benzyl chloride; C$_7$H$_7$Cl	100-44-7	126.6	Chemical ind.	Hal Arom	VOC	1.2	179	Reacts	
Biphenyl; C$_{12}$H$_{10}$	92-52-4	154.2	Chemical ind.	Arom	SVOC	8.9 × 10^{-3}	254	6.9 × 10^{-3} Insoluble	Reactive (?)[7]
Bis(2-ethylhexyl)phthalate; C$_{24}$H$_{38}$O$_4$	117-81-7	390.6	Chemical ind.; plasticizer	Phthal	SVOC	1.4 × 10^{-7}	387	3.4 × 10^{-4}	

TABLE 2.2 (CONTINUED)

Properties, Sources/Uses, and Chemical/Volatility Group Classifications of CAAA Title III HAPs (Chemicals shown in italics are high priority urban HAPs)

Compound	Chemical Formula/Structure	CAS No.	MW	Sources[1]	Chemical Class[2]	Volatility Class[3]	VP[4] (mmHG) at 25°C	BP[4]	Water Solubility[4]	Comment
Bis(chloromethyl) ether; $C_2H_4Cl_2O$	$ClCH_2OCH_2Cl$	542-88-1	115.0	Chemical ind.	Oxy Org	VOC	30.0 at 22°C	104	22 (Reacts)	Reactive[8]
Bromoform; $CHBr_3$	$CHBr_3$	75-25-2	252.7	Chemical ind.	Hal Hydro	VOC	5.4	150	3.1	
1,3-Butadiene; C_4H_6	$H_2C=CHCH=CH_2$	*106-99-0*	*54.1*	*Chemical ind.; plastics*	*Hydro*	*VVOC*	*2113*	*-4.4*	*0.5 (Insoluble)*	*Reactive (?)[7]*
Calcium cyanamide; $CaCN_2$	$CaNCN$	156-62-7	80.1	Chemical ind.	Inorg	NVINC	$<<1.0 \times 10^{-10}$	>1150	Insoluble	Reactive[5]
Captan; $C_9H_8Cl_3NO_2S$		133-06-2	300.6	Pesticide	Pestic	SVOC	9.0×10^{-8}	479	3.3×10^{-3}	Pesticide
Carbaryl; $C_{12}H_{11}NO_2$		63-25-2	201.2	Pesticide	Pestic	SVOC	1.4×10^{-6}	315	0.083	Pesticide
Carbon disulfide; CS_2	CS_2	75-15-0	76.1	Chemical ind.	Inorg	VOC	361	46	1.2	
Carbon tetrachloride; CCl_4	*CCl4*	*56-23-5*	*153.8*	*Chemical ind.*	*Hal Hydro*	*VOC*	*114*	*77*	*0.80*	
Carbonyl sulfide; COS	COS	463-58-1	60.1	Chemical ind.	Oxy Org	VVOC	9623	-50	1.2	

Name; Formula	CAS	MW	Use	Class	Type	Vapor Pressure	BP	Value	Comments
Catechol; $C_6H_6O_2$	120-80-9	110.1	Chemical ind.	Arom	VOC	0.03 at 20°C	245	461	
Chloramben; $C_7H_5Cl_2NO_2$	133-90-4	206.0	Pesticide	Pestic	SVOC	1.0×10^{-7}	312	0.8	Pesticide
Chlordane; $C_{10}H_6Cl_8$	57-74-9	409.8	Pesticide	Pestic	SVOC	9.8×10^{-6}	175 at 2 mm	5.6×10^{-5}	Pesticide - mixture of compds; VP for α- or γ-chlordane
Chlorine; Cl_2	7782-50-5	70.9	Chemical ind.; disinfectant	Inorg	VVINC	5854	-34	7.0	
Chloroacetic acid; $C_2H_3ClO_2$	79-11-8	94.5	Chemical ind.	Oxy Org	VOC	0.07	188	100 at 20°C	
2-Chloroacetophenone; C_8H_7ClO	532-27-4	154.6	Chemical ind.	Oxy Org	SVOC	5.4×10^{-3} at 20°C	245	1.6	
Chlorobenzene; C_6H_5Cl	108-90-7	112.6	Chemical ind.	Hal Arom	VOC	12.0	130	0.5	
Chlorobenzilate; $C_{16}H_{14}Cl_2O_3$	510-15-6	325.2	Chemical ind.	Pestic	SVOC	2.2×10^{-6} at 20°C	157	0.01 at 20°C	Pesticide

TABLE 2.2 (CONTINUED)
Properties, Sources/Uses, and Chemical/Volatility Group Classifications of CAAA Title III HAPs (Chemicals shown in italics are high priority urban HAPs)

Compound	Chemical Formula/Structure	CAS No.	MW	Sources[1]	Chemical Class[2]	Volatility Class[3]	VP[4] (mmHG at 25° C)	BP[4]	Water Solubility[4]	Comment
Chloroform; CHCl₃	$CHCl_3$	67-66-3	119.4	*Chemical ind.*	*Hal Hydro*	*VOC*	*197*	62	8.0	
Chloromethyl methyl ether; C₂H₅ClO	$ClCH_2OCH_3$	107-30-2	80.5	Chemical ind.	Oxy Org	VOC	187	59	69 (Reacts)	Reactive[8]
Chloroprene (2-chloro-1,3-butadiene); C₄H₅Cl	$CH_2{=}CHCCl{=}CH_2$	126-99-8	88.5	Chemical ind.; polymers	Hal Hydro	VOC	216	59	0.9 (Slightly Soluble)	
Cresols/Cresylic acid (isomer mixture); C₇H₈O	$+ CH_3, + OH$	1319-77-3	108.1	Chemical ind.; coke ovens	Arom	VOC	0.1	191-202	19.3	
o-Cresol; C₇H₈O	OH CH₃	95-48-7	108.1	Chemical ind.; coke ovens	Arom	VOC	0.2	191	25.9	
m-Cresol; C₇H₈O	OH CH₃	108-39-4	108.1	Chemical ind.; coke ovens	Arom	SVOC	0.1	202	22.7	

Compound	Structure	CAS No.	MW	Source/Use	Class	VOC Class	VP	BP	Solubility	Notes
p-Cresol; C_7H_8O		106-44-5	108.1	Chemical ind.; coke ovens	Arom	SVOC	0.1	202	<10 at 21°C	
Cumene; C_9H_{12}		98-82-8	120.2	Chemical ind.	Arom	VOC	3.5	151	0.05 (Insoluble)	
2,4-D (2,4-Dichlorophenoxyacetic acid, incl salts & esters); $C_8H_6Cl_2O_3$		N/A	221.0	Herbicide	Pestic	SVOC/ NVOC	1.0×10^{-4} to 1.0×10^{-10}	135 at 1 mm	0.9 (Slightly Soluble)	Pesticide; VP range for acid, esters, and salts; BP for acid
DDE (1,1-dichloro-2,2-bis(p-chlorophenyl) ethylene); $C_{14}H_8Cl_4$		72-55-9	318.0	Pesticide	Pestic	SVOC	2.4×10^{-5}	317	1.3×10^{-6} (Insoluble)	Pesticide
Diazomethane; CH_2N_2	CH_2N_2	334-88-3	42.0	Chemical ind.	Nitro Org	VVOC	2800	-23	2.5 (Reacts)	Highly reactive[5]
Dibenzofuran; $C_{12}H_8O$		132-64-9	168.2	Combustion products	Oxy Org	SVOC	2.5×10^{-3}	285	3.1×10^{-3} (Slightly Soluble)	Higher chlorinated species (e.g., octa) are SVOCs to NVOCs

TABLE 2.2 (CONTINUED)
Properties, Sources/Uses, and Chemical/Volatility Group Classifications of CAAA Title III HAPs (Chemicals shown in italics are high priority urban HAPs)

Compound	Chemical Formula/Structure	CAS No.	MW	Sources[1]	Chemical Class[2]	Volatility Class[3]	VP[4] (mmHG at 25°C)	BP[4]	Water Solubility[4]	Comment
1,2-Dibromo-3-chloropropane; $C_3H_5Br_2Cl$	$BrCH_2BrCHCH_2Cl$	96-12-8	236.3	Pesticide	Hal Hydro	VOC	0.6 at 20°C	195	1.2 at 20°C	
Dibutylphthalate; $C_{16}H_{22}O_4$		84-74-2	278.3	Chemical ind.	Phthal	SVOC	2.0×10^{-5}	340	0.01 (Insoluble)	
1,4-Dichlorobenzene (p-); $C_6H_4Cl_2$		106-46-7	147.0	Chemical ind.	Hal Arom	VOC	1.8	173	0.08 (Insoluble)	
3,3'-Dichlorobenzidine; $C_{12}H_{10}Cl_2N_2$		91-94-1	253.1	Chemical ind.	Nitro Org	SVOC	2.6×10^{-7}	402	0.01 (Insoluble)	
Dichloroethyl ether (Bis[2-chloroethyl]ether); $C_4H_8Cl_2O$	$(ClCH_2CH_2)_2O$	111-44-4	143.0	Chemical ind.	Oxy Org	VOC	1.6	179	17.2 at 20°C (Reacts)	Reactive (?)[7]

Name; Formula	CAS	Structure	MW	Source	Category	Class	Vapor press.	BP	Solubility	Remarks
1,3-Dichloropropene; $C_3H_4Cl_2$ *(cis)*	*542-75-6*	$CH_2ClCH{=}CHCl$	*111.0*	*Pesticide*	*Hal Hydro*	*VOC*	*34.0*	*108*	*2.8 at 20°C*	
Dichlorvos; $C_4H_7Cl_2O_4P$	62-73-7	CH_3O, CH_3O $P{-}O{-}CH{=}CCl_2$ (with $\overset{O}{\underset{\|}{}}$)	221.0	Pesticide	Pesticide	SVOC	1.6×10^{-2}	234	16.0	Pesticide
Diethanolamine; $C_4H_{11}NO_2$	111-42-2	$(HOC_2H_4)_2NH$	105.1	Chemical ind.	Nitro Org	SVOC	2.5×10^{-4}	269	1000 at 20°C	Reactive (?)[7]; strong base
Diethyl sulfate; $C_4H_{10}O_4S$	64-67-5	$(C_2H_5)_2SO_4$	154.2	Chemical ind.	Sulfat	VOC	0.21	208	7.0 at 20°C (Reacts)	Reactive (?)[7]
3,3'-Dimethoxybenzidine; $C_{14}H_{16}N_2O_2$	119-90-4	(structure with NH_2, OCH_3, H_2N, CH_3O)	244.3	Chemical ind.	Nitro Org	NVOC	1.25×10^{-7}	356	<0.1 at 20°C	
4-Dimethylamino-azobenzene; $C_{14}H_{15}N_3$	60-11-7	(structure with CH_3, CH_3 N, $N{=}N$)	225.3	Chemical ind.	Nitro Org	NVOC	7.0×10^{-8}	407	2.3×10^{-4}	
N,N-Dimethylaniline; $C_8H_{11}N$	121-69-7	(structure CH_3, $N{-}CH_3$)	121.2	Chemical ind.	Nitro Org	VOC	0.8	194	1.5	
3,3'-Dimethylbenzidine; $C_{14}H_{16}N_2$	119-93-7	(structure with NH_2, CH_3, H_2N, H_3C)	212.3	Chemical ind.	Nitro Org	SVOC	6.9×10^{-7}	300	1.3	
Dimethyl carbamoyl chloride; C_3H_6ClNO	79-44-7	$(CH_3)_2NC(O)Cl$	107.5	Chemical ind.	Nitro Org	VOC	2.0	166	Reacts	Highly reactive[8]

TABLE 2.2 (CONTINUED)
Properties, Sources/Uses, and Chemical/Volatility Group Classifications of CAAA Title III HAPs (Chemicals shown in italics are high priority urban HAPs)

Compound	Chemical Formula/Structure	CAS No.	MW	Sources[1]	Chemical Class[2]	Volatility Class[3]	VP[4] (mmHG) at 25° C	BP[4]	Water Solubility[4]	Comment
N,N-Dimethylformamide; C_3H_7NO	$HC(O)N(CH_3)_2$	68-12-2	73.1	Chemical ind.	Nitro Org	VOC	2.6	153	1000	
1,1-Dimethylhydrazine; $C_2H_8N_2$	$(CH_3)_2NNH_2$	57-14-7	60.1	Chemical ind.; rocket fuel	Nitro Org	VOC	157	63	1000 (Reacts)	Reactive (?)[7]
Dimethyl phthalate; $C_{10}H_{10}O_4$		131-11-3	194.2	Chemical ind.; plasticizer	Phthal	SVOC	3.1×10^{-3}	284	4.2	
Dimethyl sulfate; $C_2H_6O_4S$	$(CH_3)_2SO_4$	77-78-1	126.1	Chemical ind.	Sulfate	VOC	0.7	189	28.0 at 18°C	Reactive (?)[7]
4,6-Dinitro-o-cresol & salts; $C_7H_6N_2O_5$		N/A	198.1	Pesticide	Nitro Org	SVOC	1.1×10^{-4}	312	0.1 (Slightly Soluble)	Pesticide; VP, BP for the cresol; salts are probably NVOCs

Compound; Formula	CAS No.	MW	Use	Class	VOC/SVOC	Vapor pressure	B.P.	Solubility	Reactivity
2,4-Dinitrophenol; $C_6H_4N_2O_5$	51-28-5	184.1	Chemical ind.	Nitro Org	SVOC	3.9×10^{-4} at 20°C	Sublimes on heating	2.8	
2,4-Dinitrotoluene; $C_7H_6N_2O_4$	121-14-2	182.1	Chemical ind.	Nitro Org	SVOC	2.4×10^{-4}	300	0.3	
1,4-Dioxane (1,4-Diethylene oxide); $C_4H_8O_2$	123-91-1	88.1	Chemical ind.	Oxy Org	VOC	27.0	101	1000	
1,2-Diphenylhydrazine; $C_{12}H_{12}N_2$	122-66-7	184.2	Chemical ind.	Nitro Org	SVOC	4.4×10^{-4}	309	0.07 (Insoluble)	Reactive (?)[7]
Epichlorohydrin (1-chloro-2,3-epoxy propane); C_3H_5ClO	106-89-8	92.5	Chemical ind.	Oxy Org	VOC	12.5	118	65.9	Highly reactive[6]

TABLE 2.2 (CONTINUED)
Properties, Sources/Uses, and Chemical/Volatility Group Classifications of CAAA Title III HAPs (Chemicals shown in italics are high priority urban HAPs)

Compound	Chemical Formula/Structure	CAS No.	MW	Sources[1]	Chemical Class[2]	Volatility Class[3]	VP[4] (mmHG) at 25° C	BP[4]	Water Solubility[4]	Comment
1,2-Epoxybutane; C_4H_8O	$H_2C\!-\!CHCH_2CH_3$ (with O epoxide)	106-88-7	72.1	Chemical ind.; gasoline additive	Oxy Org	VOC	180	63	95.0	Reactive[6]
Ethyl acrylate; $C_5H_8O_2$	$H_2C=CHCO_2C_2H_5$	140-88-5	100.1	Chemical ind.	Oxy Org	VOC	38.4	100	15.0	
Ethylbenzene; C_8H_{10}	C_2H_5 (benzene ring)	100-41-4	106.2	Chemical ind.	Arom	VOC	12.7	136	0.2	
Ethyl carbamate (urethane); $C_3H_7NO_2$	$NH_2C(O)OC_2H_5$	51-79-6	89.1	Chemical ind.	Nitro Org	VOC	0.3	182	480 at 15°C	
Ethyl chloride; C_2H_5Cl	CH_3CH_2Cl	75-00-3	64.5	Chemical ind.	Hal Hydro	VVOC	1202	12	5.7 at 20°C	
Ethylene dibromide; $C_2H_4Br_2$ (1,2-Dibromoethane)	CH_2BrCH_2Br	*106-93-4*	*187.9*	*Chemical ind.*	*Hal Hydro*	*VOC*	*14.3*	*132*	*4.2*	*Pesticide*
Ethylene dichloride; $C_2H_4Cl_2$ (1,2-Dichloroethane);	CH_2ClCH_2Cl	*107-06-2*	*99.0*	*Chemical ind.*	*Hal Hydro*	*VOC*	*78.7*	*84*	*8.6*	*Pesticide*
Ethylene glycol; $C_2H_6O_2$	$HOCH_2CH_2OH$	107-21-1	62.1	Chemical ind.; antifreeze	Oxy Org	SVOC	0.06 at 20°C	195	1000	

Compound	Structure	CAS	MW	Sources	Class	Volatility				Notes
Ethylene imine; C_2H_5N		151-56-4	43.1	Chemical ind.	Nitro Org	VOC	213	56	1000 (Miscible)	Reactive (?)[7]
Ethylene oxide; C_2H_4O		*75-21-8*	*44.1*	*Chemical ind.; hospital sterilizers*	*Oxy Org*	*VVOC*	*1311*	*11*	*1000 (Miscible)*	*Reactive[6]*
Ethylene thiourea; $C_3H_6N_2S$		96-45-7	102.2	Chemical ind.	Nitro Org	SVOC	2.0×10^{-6}	347	1-5 at 18°C	
Ethylidene dichloride (1,2-Dichloroethane); $C_2H_4Cl_2$	CH_3CHCl_2	75-34-3	99.0	Chemical ind.	Hal Hydro	VOC	227	57	5.1	
Formaldehyde; CH_2O	*HCHO*	*50-00-0*	*30.0*	*Chemical ind.; photochemical reaction prod.; combustion sources*	*Oxy Org*	*VVOC*	*3821*	*-20*	*550*	
Heptachlor; $C_{10}H_5Cl_7$		76-44-8	373.3	Pesticide	Pesticide	SVOC	4.0×10^{-4}	145 at 1.5 mm	0.002	Pesticide

TABLE 2.2 (CONTINUED)
Properties, Sources/Uses, and Chemical/Volatility Group Classifications of CAAA Title III HAPs (Chemicals shown in italics are high priority urban HAPs)

Compound	Chemical Formula/Structure	CAS No.	MW	Sources[1]	Chemical Class[2]	Volatility Class[3]	VP[4] (mmHG) at 25°C	BP[4]	Water Solubility[4]	Comment
Hexachlorobenzene; C_6Cl_6		*118-74-1*	*284.8*	*Pesticide*	*Hal Arom*	*SVOC*	*5.2×10^{-3}*	*332*	*6.2×10^{-6}*	*Pesticide*
Hexachlorobutadiene; C_4Cl_6	$Cl_2C=CCl-CCl=CCl_2$	87-68-3	260.8	Chemical ind.	Hal Hydro	VOC	0.2	210	3.2×10^{-3}	
1,2,3,4,5,6-Hexachlorocyclohexane and isomers (e.g., Lindane 58-89-9); $C_6H_6Cl_6$		N/A	290.8	Pesticide	Pestic	SVOC	4.2×10^{-5} at 20°C	323	7.3×10^{-3}	Pesticide
Hexachlorocyclopentadiene; C_5Cl_6		77-47-4	272.8	Pesticide	Hal Hydro	SVOC	6.0×10^{-2}	239	3.4×10^{-3}	Reactive (?)[7]
Hexachloroethane; C_2Cl_6	Cl_3CCCl_3	67-72-1	236.7	Chemical ind.	Hal Hydro	VOC	0.6	Sublimes at 189°C	0.05	
Hexamethylene-1,6-diisocyanate; $C_8H_{12}N_2O_2$	$OCN(CH_2)_6NCO$	822-06-0	168.2	Chemical ind.	Nitro Org	SVOC	0.03	255	0.1	Reactive (?)[7]

Name; Formula	Structure	CAS	MW	Use	Class	Volatility			
Hexamethylphosphoramide; $C_6H_{18}N_3OP$	$[(CH_3)_2N]_3PO$	680-31-9	179.2	Chemical ind.	Nitro Org	SVOC	0.05	235	1000
Hexane; C_6H_{14}	$CH_3(CH_2)_4CH_3$	110-54-3	86.2	Chemical ind.; petroleum	Hydro	VOC	151	69	9.5×10^{-3}
Hydrazine; H_4N_2	*H2N-NH2*	*302-01-2*	*32.0*	*Chemical ind.; fuel*	*Nitro Org*	*VINC*	*14.4*	*114*	*1000 (Miscible)*
Hydrogen chloride; HCl	HCl	7647-01-0	36.5	Chemical ind.	Inorg	VINC	35771	-85	620
Hydrogen fluoride; HF	HF	7664-39-3	20.0	Chemical ind.	Inorg	VVINC	905	20	Very soluble
Hydroquinone; $C_6H_6O_2$	OH – (ring) – OH	123-31-9	110.1	Chemical ind.	Oxy Org	SVOC	8.6×10^{-4}	285	70
Isophorone; $C_9H_{14}O$	(structure)	78-59-1	138.2	Chemical ind.; solvent	Oxy Org	VOC	0.4	215	12
Maleic anhydride; $C_4H_2O_3$	(structure)	108-31-6	98.1	Chemical ind.	Oxy Org	SVOC	0.3	200	4.9 (Soluble) Reactive[5]

TABLE 2.2 (CONTINUED)
Properties, Sources/Uses, and Chemical/Volatility Group Classifications of CAAA Title III HAPs (Chemicals shown in italics are high priority urban HAPs)

Compound	Chemical Formula/Structure	CAS No.	MW	Sources[1]	Chemical Class[2]	Volatility Class[3]	VP[4] (mmHG) at 25° C	BP[4]	Water Solubility[4]	Comment
Methanol; CH_4O	CH_3OH	67-56-1	32.0	Chemical ind.; fuel	Oxy Org	VOC	118	65	1000	
Methoxychlor; $C_{16}H_{15}Cl_3O_2$		72-43-5	345.7	Pesticide	Pestic	SVOC	2.6×10^{-6}	346	1.0×10^{-4}	Pesticide
Methyl bromide; CH_3Br	CH_3Br	74-83-9	94.9	Pesticide	Hal Hydro	VVOC	1646	4	15.2	Pesticide
Methyl chloride; CH_3Cl	CH_3Cl	74-87-3	50.5	Chemical ind.	Hal Hydro	VVOC	4318	-24	5.3	
Methyl chloroform (1,1,1-trichloroethane); $C_2H_3Cl_3$	CCl_3CH_3	71-55-6	133.4	Chemical ind.; degreasing solvent	Hal Hydro	VOC	133	74	1.5	
Methyl ethyl ketone (2-butanone); C_4H_8O	$C_2H_5COCH_3$	78-93-3	72.1	Chemical ind.; solvent	Oxy Org	VOC	90.0	80	256 at 20°C	
Methylhydrazine; CH_6N_2	CH_3NHNH_2	60-34-4	46.1	Chemical ind.	Nitro Org	VOC	49.6	88	1000	Highly reactive[8]
Methyl iodide (iodomethane); CH_3I	CH_3I	74-88-4	141.9	Chemical ind.	Hal Hydro	VVOC	402	42	1.4	
Methyl isobutyl ketone (hexone); $C_6H_{12}O$	$(CH_3)_2CHCH_2COCH_3$	108-10-1	100.2	Chemical ind.; solvent	Oxy Org	VOC	19.7	117	19	

Name; formula	Structure	CAS	MW	Source/use	Class	VOC type	Vapor pressure	b.p.	Solubility	Reactivity
Methyl isocyanate; C_2H_3NO	CH_3NCO	624-83-9	57.1	Chemical ind.	Nitro Org	VOC	348 at 20°C	60	67 (Reacts)	Highly reactive[8]
Methyl methacrylate; $C_5H_8O_2$	$H_2C=C(CH_3)CO_2CH_3$	80-62-6	100.1	Chemical ind.	Oxy Org	VOC	36.1	100	15	
Methyl-tert-butyl ether; $C_5H_{12}O$	$(CH_3)_3COCH_3$	1634-04-4	88.1	Petroleum; gasoline additive	Oxy Org	VOC	250	55	51	
4,4'-Methylenebis(2-chloroaniline); $C_{13}H_{12}Cl_2N_2$	[structure]	101-14-4	267.2	Chemical ind.	Nitro Org	NVOC	2.9×10^{-7}	379	0.01 at 24°C	
Methylene chloride; CH_2Cl_2 (dichloromethane)	*Cl-CH2-Cl*	75-09-2	84.9	*Chemical ind.; paint removal solvent*	*Hal Hydro*	*VOC*	*435*	*40*	*13.2*	
4,4'-Methylenediphenyl diisocyanate (MDI); $C_{15}H_{10}N_2O_2$	[structure: OCN–CH2–NCO]	101-68-8	250.3	Chemical ind.	Nitro Org	SVOC	1.2×10^{-5}	194 at 5 mm	0.83×10^{-3} (Insoluble)	Reactive (?)[7]
4,4'-Methylenedianiline; $C_{13}H_{14}N_2$	[structure: H_2N–CH2–NH_2]		198.3	Chemical ind.	Nitro Org	NVOC	1.7×10^{-10}	398	<1 at 19°C	
Naphthalene; $C_{10}H_8$	[structure]	91-20-3	128.2	Chemical ind.; coke ovens	Arom	SVOC	0.26	218	0.03	
Nitrobenzene; $C_6H_5NO_2$	[structure: NO_2 benzene]	98-95-3	123.1	Chemical ind.	Nitro Org	VOC	0.25	211	1.9	
4-Nitrobiphenyl; $C_{12}H_9NO_2$	[structure: NO_2 biphenyl]	92-93-3	199.2	Chemical ind.	Nitro Org	SVOC	3.3×10^{-5}	340	9.8×10^{-3}	

TABLE 2.2 (CONTINUED)
Properties, Sources/Uses, and Chemical/Volatility Group Classifications of CAAA Title III HAPs (Chemicals shown in italics are high priority urban HAPs)

Compound	Chemical Formula/Structure	CAS No.	MW	Sources[1]	Chemical Class[2]	Volatility Class[3]	VP[4] (mmHG at 25° C)	BP[4]	Water Solubility[4]	Comment
4-Nitrophenol; $C_6H_5NO_3$	NO_2 — OH (ring structure)	100-02-7	139.1	Chemical ind.	Nitro Org	SVOC	4.1×10^{-5} at 20°C	279	16.0	
2-Nitropropane; $C_3H_7NO_2$	$CH_3CH(NO_2)CH_3$	79-46-9	89.1	Chemical ind.	Nitro Org	VOC	17.0	120	17.0	
N-Nitroso-N-methylurea; $C_2H_5N_3O_2$	$CH_3N(NO)C(O)NH_2$	684-93-5	103.1	Chemical ind.	Nitro Org	VOC	0.03	124	14.4 at 23°C	Reactive[8]
N-Nitrosodimethylamine; $C_2H_6N_2O$	$(CH_3)_2NNO$	62-75-9	74.1	Chemical ind.	Nitro Org	VOC	2.7 at 20°C	149	1000 at 24°C	Reactive[8]
N-Nitrosomorpholine; $C_4H_8N_2O_2$	$N=\dot{O}$ (morpholine structure)	59-89-2	116.1	Chemical ind.	Nitro Org	VOC	0.04 at 20°C	224	1000 at 24°C	

Name; Formula	CAS	Structure	MW	Use	Class	Category	Vapor pressure			Use
Parathion; $C_{10}H_{14}NO_5PS$	56-38-2		291.3	Pesticide	Pestic	SVOC	6.7×10^{-6} at 20°C	375	6.5×10^{-3}	Pesticide
Pentachloronitrobenzene (quintobenzene); $C_6Cl_5NO_2$	82-68-8		295.3	Chemical ind.	Nitro Org	SVOC	5.0×10^{-5} at 20°C	328	5.9×10^{-4}	
Pentachlorophenol; C_6Cl_5OH	87-86-5		266.3	Pesticide	Oxy Org	SVOC	1.1×10^{-4}	310	0.014	Pesticide
Phenol; C_6H_6O	108-95-2		94.1	Chemical ind.	Oxy Org	VOC	0.35	182	82.8	
p-Phenylenediamine; $C_6H_8N_2$	106-50-3		108.1	Chemical ind.	Nitro Org	SVOC	5.0×10^{-3}	267	47.0	

TABLE 2.2 (CONTINUED)
Properties, Sources/Uses, and Chemical/Volatility Group Classifications of CAAA Title III HAPs (Chemicals shown in italics are high priority urban HAPs)

Compound	Chemical Formula/Structure	CAS No.	MW	Sources[1]	Chemical Class[2]	Volatility Class[3]	VP[4] (mmHG) at 25° C	BP[4]	Water Solubility[4]	Comment
Phosgene; CCl_2O	$COCl_2$	75-44-5	98.9	Chemical ind.	Oxy Org	VVOC	1406	8	Slightly Soluble	Reactive (?)[7]
Phosphine; PH_3	PH_3	7803-51-2	34.0	Chemical ind.	Inorg	VVINC	2.84×10^4	-87	0.4 (Slightly Soluble)	Pesticide
Phosphorus; P	P_4	7723-14-0	124.0	Chemical ind.	Inorg	SVINC	4.6×10^{-2}	280	0.003	Highly reactive (red and white forms)
Phthalic anhydride; $C_8H_4O_3$		85-44-9	148.1	Chemical ind.	Phthal	SVOC	9.6×10^{-4}	284	6.2 (Reacts)	Reactive[5]
Polychlorinated biphenyls (PCBs):	*Biphenyl with Cl at various levels of saturation*	*1336-36-3*		*Dielectrics (no longer used)*	*Hal Arom*					
2-Chlorobiphenyl; $C_{12}H_9Cl$		2051-60-7	188.7			SVOC	1.2×10^{-2}	274	4.8×10^{-3}	Higher chlorinated species (up to deca-) are NVOCs

Name; Formula	Structure	CAS No.	MW	Source	Class	VOC Class				Notes
2,4,6-Trichlorobiphenyl; $C_{12}H_7Cl_3$		35693-92-6	257.5			SVOC	7.2×10^{-4}	423	2.5×10^{-4}	Reactive (?)[7]
1,3-Propane sultone; $C_3H_6O_3S$		1120-71-4	122.1	Chemical ind.	Oxy Org	VOC	0.27	112 at 30 mm	171	
β-Propiolactone; $C_3H_4O_2$		57-57-8	72.1	Chemical ind.	Oxy Org	VOC	3.4	Decomposes at 155°C	10-50 at 19°C	
Propionaldehyde; C_3H_6O	C_2H_5CHO	123-38-6	58.1	Chemical ind.	Oxy Org	VOC	317	49	306	
Propoxur (Baygon); $C_{11}H_{15}NO_3$		114-26-1	209.2	Pesticide	Pestic	SVOC	9.7×10^{-6} at 20°C	400	2.0 (Slightly Soluble)	Pesticide
Propylene dichloride; $C_3H_6Cl_2$ (1,2-dichloropropane)	$CH_3CHClCH_2Cl$	78-87-5	113.0	*Chemical ind.*	*Hal Hydro*	*VOC*	*50.4*	97	2.7	*Pesticide*
Propylene oxide; C_3H_6O		75-56-9	58.1	Chemical ind.	Oxy Org	VVOC	530	34	590	Reactive[6]
1,2-Propyleneimine; C_3H_7N (2-methylaziridine)		75-55-8	57.1	Chemical ind.	Nitro Org	VOC	112 at 20°C	66	1000	Highly Reactive (?)[7]

TABLE 2.2 (CONTINUED)
Properties, Sources/Uses, and Chemical/Volatility Group Classifications of CAAA Title III HAPs (Chemicals shown in italics are high priority urban HAPs)

Compound	Chemical Formula/Structure	CAS No.	MW	Sources[1]	Chemical Class[2]	Volatility Class[3]	VP[4] (mmHG at 25° C)	BP[4]	Water Solubility[4]	Comment
Quinoline; C₉H₇N		*91-22-5*	*129.2*	*Chemical ind.*	*Nitro Org*	*SVOC*	*8.3 × 10⁻²*	*238*	*<0.1 at 22.5°C*	
Quinone; C₆H₄O₂		106-51-4	108.1	Chemical ind.	Oxy Org	SVOC	9.0×10^{-2}	201	Slightly Soluble	
Styrene; C₈H₈		100-42-5	104.2	Chemical ind.	Arom	VOC	6.1	145	0.32	
Styrene oxide; C₈H₈O		96-09-3	120.2	Chemical ind.	Oxy Org	VOC	0.3 at 20°C	194	3.0 at 20°C	Highly reactive[8]
2,3,7,8-Tetrachlorodibenzo-p-dioxin; C₁₂H₄Cl₄O₂		*1746-01-6*	*322.0*	*Combustion product*	*Hal Arom*	*SVOC*	*1.5 × 10⁻⁹*	*500*	*2.0 × 10⁻⁸*	
1,1,2,2-Tetrachloroethane; C₂H₂Cl₄	$Cl_2CHCHCl_2$	*79-34-5*	*167.8*	*Chemical ind.*	*Hal Hydro*	*VOC*	*4.0*	*146*	*3.0*	

Name; Formula	Structure	CAS	MW	Use/Source	Class	Volatility				Reactivity/Notes
Tetrachloroethylene; C_2Cl_4 (perchloroethylene)	$Cl_2C{=}CCl_2$	*127-18-4*	*165.8*	*Chemical ind.; dry cleaning*	*Hal Hydro*	*VOC*	*18.6*	*121*	*0.2*	
Titanium tetrachloride; $TiCl_4$	$TiCl_4$	7550-45-0	189.7	Chemical ind.	Inorg	VINC	12.4	136	Soluble	Highly reactive in air; forms TiO_2
Toluene; C_7H_8	CH_3 (benzene ring)	108-88-3	92.1	Petroleum; solvent; gasoline	Arom	VOC	28.6	111	0.53	
2,4-Toluene–diamine; $C_7H_{10}N_2$	CH_3, NH_2, NH_2 (benzene ring)	95-80-7	122.2	Chemical ind.	Nitro Org	SVOC	2.3×10^{-3}	292	75.0	
2,4-Toluene diisocyanate; $C_9H_6N_2O_2$	CH_3, OCN, OCN (benzene ring)	584-84-9	174.2	Chemical ind.	Nitro Org	SVOC	0.016	251	0.038 (Reacts)	Reactive[5]
o-Toluidine; C_7H_9N	CH_3, NH_2 (benzene ring)	95-53-4	107.2	Chemical ind.	Nitro Org	SVOC	0.25	200	16.6	

TABLE 2.2 (CONTINUED)
Properties, Sources/Uses, and Chemical/Volatility Group Classifications of CAAA Title III HAPs (Chemicals shown in italics are high priority urban HAPs)

Compound	Chemical Formula/Structure	CAS No.	MW	Sources[1]	Chemical Class[2]	Volatility Class[3]	VP[4] (mmHG at 25° C)	BP[4]	Water Solubility[4]	Comment
Toxaphene (chlorinated camphene); $C_{10}H_{10}Cl_8$		8001-35-2	414 (ave.)	Pesticide	Pestic	SVOC	1.1×10^{-5}	155 at 0.4 mm	5.5×10^{-4}	Pesticide; complex mixture of isomers
1,2,4-Trichlorobenzene; $C_6H_3Cl_3$		120-82-1	181.4	Chemical ind.	Hal Arom	VOC	0.42	214	0.049	
1,1,2-Trichloroethane; $C_2H_3Cl_3$	$CH_2ClCHCl_2$	79-00-5	133.4	Chemical ind.	Hal Hydro	VOC	22.0	114	4.4	
Trichloroethylene; C_2HCl_3	$Cl_2C{=}CHCl$	79-01-6	131.4	Chemical ind.; degreasing solvent	Hal Hydro	VOC	69.0	87	1.1	

Name	Structure	CAS	MW	Use	Class	Volatility	VP	BP	Sol	Notes
2,4,5-Trichlorophenol; $C_6H_3Cl_3O$		95-95-4	197.4	Herbicide	Oxy Org	SVOC	4.0×10^{-2}	253	1.2	Pesticide
2,4,6-Trichlorophenol; $C_6H_3Cl_3O$		88-06-2	197.4	Herbicide	Oxy Org	SVOC	2.5×10^{-2}	245	0.8	Pesticide
Triethylamine; $C_6H_{15}N$	$(C_2H_5)_3N$	121-44-8	101.2	Chemical ind.	Nitro Org	VOC	57.9	89	74.0 (Soluble)	Reactive (?)[7]; strong base
Trifluralin; $C_{13}H_{16}F_3N_3O_4$		1582-09-8	335.3	Chemical ind.	Pestic	SVOC	4.6×10^{-5}	139 at 4.2 mm	1.8×10^{-4}	Pesticide
2,2,4-Trimethyl pentane; C_8H_{18}	$CH_3C(CH_3)_2CH_2CH(CH_3)CH_3$	540-84-1	114.2	Chemical ind.; petroleum	Hydro	VOC	48.7	99	2.4×10^{-3}	
Vinyl acetate; $C_4H_6O_2$	$CH_3CO_2CH{=}CH_2$	108-05-4	86.1	Chemical ind.	Oxy Org	VOC	115	72	20.0 at 20°C	
Vinyl bromide (bromoethene); C_2H_3Br	$H_2C{=}CHBr$	593-60-2	106.9	Chemical ind.	Hal Hydro	VVOC	1059	16	5.7	
Vinyl chloride; C_2H_3Cl (chloroethene);	$H_2C{=}CHCl$	75-01-4	62.5	Chemical ind.; polymers	Hal Hydro	VVOC	2937	-14	1.1	
Vinylidene chloride; $C_2H_2Cl_2$ (1,1-dichloroethylene)	$H_2C{=}CCl_2$	75-35-4	96.9	Chemical ind.; polymers	Hal Hydro	VVOC	600	32	2.3	

TABLE 2.2 (CONTINUED)
Properties, Sources/Uses, and Chemical/Volatility Group Classifications of CAAA Title III HAPs (Chemicals shown in italics are high priority urban HAPs)

Compound	Chemical Formula/Structure	CAS No.	MW	Sources[1]	Chemical Class[2]	Volatility Class[3]	VP[4] (mmHG) at 25° C	BP[4]	Water Solubility[4]	Comment
Xylenes (isomer mixture); C_8H_{10}	(structure)	1330-20-7	106.2	Petroleum; solvent; gasoline	Arom	VOC	8.0	140	0.2	
o-Xylene; C_8H_{10}	(structure)	95-47-6	106.2	Petroleum; solvent; gasoline	Arom	VOC	6.6	144	0.2	
m-Xylene; C_8H_{10}	(structure)	108-38-3	106.2	Petroleum; solvent; gasoline	Arom	VOC	8.3	139	0.2	
p-Xylene; C_8H_{10}	(structure)	106-42-3	106.2	Petroleum; solvent; gasoline	Arom	VOC	8.9	138	0.2	

Compound	Description	CAS	MW	Source	Class	Volatility	VP	MP	Solubility	Notes
Antimony Compounds	Inorganic compounds in particle phase	--	--	Metals; chemicals	Inorg	NVINC	1 mm at 574°C	655 (mp)	Insoluble	VP and MP for antimony trioxide[4c]; Volatile forms exist, e.g., stilbene, SbH_3
Arsenic compounds:	Primarily inorganic arsenic compounds in particle phase			Smelting; chemicals; pesticides	Inorg					
e.g., Arsine, AsH_3	AsH_3	7784-42-1	77.9			VVINC	10.2×10^3	-55	20 mL/100 mL water at 20°C	
e.g., particle phase	Inorganic compounds in particle phase	--	--			NVINC	1 mm at 212°C	313 (mp)	Insoluble	VP and MP given for arsenic trioxide[4c]
Beryllium compounds	Inorganic compounds in particle phase	--	--	Metals; ceramics	Inorg	NVINC			Insoluble	
Cadmium compounds	Inorganic compounds in particle phase	--	--	Chemicals; smelting	Inorg	NVINC	1 mm at 1000°C		Insoluble	VP given for cadmium oxide[4c]
Chromium compounds	Inorganic compounds in particle phase	--	--	Metals; plating	Inorg	NVINC			Insoluble	Semivolatile forms can also exist in air, e.g., $Cr(CO)_6$
Cobalt compounds	Inorganic compounds in particle phase	--	--	Metals	Inorg	NVINC			Insoluble	Semivolatile forms can also exist in air, e.g., $Co(CO)_4$
Coke Oven Emissions	Mixture of organic and inorganic vapors and particles			Coke ovens; steel	Arom					Emissions include VOCs, e.g., benzene, toluene, xylenes

TABLE 2.2 (CONTINUED)
Properties, Sources/Uses, and Chemical/Volatility Group Classifications of CAAA Title III HAPs (Chemicals shown in italics are high priority urban HAPs)

Compound	Chemical Formula/Structure	CAS No.	MW	Sources[1]	Chemical Class[2]	Volatility Class[3]	VP[4] (mmHG at 25°C)	BP[4]	Water Solubility[4]	Comment
e.g., Naphthalene; $C_{10}H_8$		91-20-3	128.2			SVOC	0.26	218	0.03	
e.g., Coronene; $C_{24}H_{12}$		191-07-1	300.4			NVOC	3.3×10^{-11}	525	1.4×10^{-7}	
Cyanide Compounds	e.g., HCN (74-90-8), propionitrile, C_2H_5CN (107-12-0), cyanogen, C_2N_2 (74-87-5). (See also acetonitrile, acrylonitrile)			Chemical ind.	Inorg					
e.g., Hydrogen Cyanide	HCN	74-90-8	27.0			VVINC	742	26	Soluble	
e.g., particle phase		--	--			NVINC			Insoluble	
Glycol ethers	e.g., $HOCH_2CH_2OCH_3$	—	76.1 to 206.3	Chemical ind.	Oxy Org	SVOC/ VOC	0.022 to 10.9	120 to 249	10-100 at 22°C	
Lead compounds	*Primarily inorganic compounds in particle phase. Trace amounts of organo-lead compounds, e.g., tetraethyl lead $(C_2H_5)_4Pb$ (78-00-2).*	--	--	*Chemical ind.; metals*	*Inorg*	*NVINC*	*1 mm at 943°C*	*890 (mp)*	*Insoluble*	*VP and MP given for lead oxide[4c]*

Compound	Description / Structure	CAS	Mol. wt.	Source	Class	Volatility	Vapor pressure	B.P.	Solubility	Notes
Manganese compounds	Inorganic compounds in the particle phase	--	--	Metals; chemicals	Inorg	NVINC			Insoluble	
Mercury compounds	Primarily elemental mercury vapor, Hg. Small amounts of particulate inorganic Hg compounds. Gaseous organic Hg compounds, e.g., (CH3)2Hg (593-74-8)			Chlor-alkali; biogenic	Inorg					
e.g., Mercury Vapor	Hg	7439-97-6	200.6			SVINC	2.0×10^{-3} mm at 20°C	356	6.0×10^{-5}	
e.g., particle phase		--	--			NVINC			Insoluble	
Fine mineral fibers (incl. glass)	Fibrous glass, mineral wool, or ceramic fibers, similar in shape but chemically and physically distinct from asbestos fibers	--	--	Insulation	Inorg	NVINC			Insoluble	
Nickel compounds	Inorganic compounds in the particle phase	--	--	Metals	Inorg	NVINC			Insoluble	
Polycyclic Organic Matter (POM):	e.g., Phenanthrene (85-01-8), anthracene (120-12-7). (See also naphthalene)			Combustion products	Arom					Volatile forms can also exist briefly in air, e.g., Ni(CO)4
e.g., Naphthalene; $C_{10}H_8$		91-20-3	128.2			SVOC	0.26	218	0.03	PAH
e.g., Coronene; $C_{24}H_{12}$		191-07-1	300.4			NVOC	3.3×10^{-11}	525	1.4×10^{-7}	PAH

TABLE 2.2 (CONTINUED)
Properties, Sources/Uses, and Chemical/Volatility Group Classifications of CAAA Title III HAPs (Chemicals shown in italics are high priority urban HAPs)

Compound	Chemical Formula/Structure	CAS No.	MW	Sources[1]	Chemical Class[2]	Volatility Class[3]	VP[4] (mmHG at 25° C	BP[4]	Water Solubility[4]	Comment
Radionuclides	Radon gas. Radioactive compounds in the particle phase			Nuclear ind.	Inorg					
e.g., various (particle and gaseous)		10043-92-2	--			NVINC				Reactive (?)[7]
Radon; Rn-222		14859-67-7	222.0			VVINC	--	-62	224 cc	Noble gas
Selenium compounds: particulate	Inorganic compounds in the particle phase	--	--	Chemical ind.; metals	Inorg	NVINC	1 mm at 157°C	340 (mp)	Insoluble	VP and MP given for selenium dioxide[4c]

1. Data on sources from Ref. 1
2. Data on chemical class classifications from Ref. 5.
3. Compound Classes:

   VVOC = Very Volatile Organic Compounds (Vapor Pressure at 25° C >380 mm Hg)

   VVINC or Gases = Very Volatile Inorganic Compounds (Vapor Pressure at 25° C >380 mm Hg)

   VOC = Volatile Organic Compounds (1.0E-01< Vapor Pressure at 25° C <380 mm Hg)

   VINC or Gases = Volatile Inorganic Compounds (1.0E-01< Vapor Pressure at 25° C <380 mm Hg)

   SVOC = Semivolatile Organic Compounds (1.0E-07< Vapor Pressure at 25° C <1.0E-01 mm Hg)

   SVINC = Semivolatile Inorganic Compounds (1.0E-07< Vapor Pressure at 25°C <1.0E-01 mm Hg)

   NVOC = Nonvolatile Organic Compounds (Vapor Pressure at 25° C < 1.0E-07 mm Hg)

   NVINC = Non-Volatile Inorganic Compounds (Vapor Pressure at 25° C < 1.0E-07 mm Hg)

TABLE 2.2 (CONTINUED)
Properties, Sources/Uses, and Chemical/Volatility Group Classifications of CAAA Title III HAPs (Chemicals shown in italics are high priority urban HAPs)

4. Data on vapor pressure (VP), boiling point (BP)/melting point (MP), and solubility were obtained from several literature and Internet-based sources. In many cases (e.g., boiling point), values listed indicate a general agreement between sources consulted so that a single value could easily be selected. In other cases, (e.g., water solubility), literature values varied widely and the value selected was based on prior experience and personal judgment. Data were taken from: (a) Ref. 16; (b) Ref. 17; (c) R.C. Weast, Ed., *CRC Handbook of Chemistry and Physics*, 59th ed., CRC, Boca Raton, 1979; (d) Ref. 14; (e) Ref. 18; (f) Ref. 19; (g) Ref. 20;

5. *The Merck Index*, 11th ed., S. Budavari, Ed., Merck & Co., Inc., Rahway, NJ, 1989.

6. R.T. Morrison and R.N. Boyd, *Organic Chemistry*, 2nd ed., Allyn and Bacon, Inc., Boston, MA, 1966.

7. Reactive (?) or Highly Reactive (?) indicates judgment based on properties, personal communication from Robert G. Lewis, US EPA, March 1994.

8. From reactivity data in Table 2.3.

3 Measurement Methods for the 188 Hazardous Air Pollutants in Ambient Air

3.1 INTRODUCTION

The goals of the 1990 Clean Air Act Amendments[1] (CAAA) require measurements of HAPs in two broad complementary target areas. One is the determination of emissions of HAPs from industrial sources. Such measurements are valuable in determining emission inventories of HAPs, in establishing the category designation (i.e., major or minor) of industrial sources, in determining the impact of modifications to sources and in assessing the adequacy of emission control devices. These source-related measurements can be made by a variety of methods, at emission points, in emission plumes, or at the boundaries of industrial facilities. However, source-related measurements of HAPs do not directly address the widespread population exposure that results from the presence of HAPs in air. Dispersion modeling can be used, with measured HAPs emission rates, to estimate HAPs levels in air in communities near industrial facilities, or in a larger urban area. However, such modeling may not accurately reflect the transport and transformation of HAPs in air or adequately include additional emissions of HAPs from the numerous small emitters collectively called "area sources."

To assess the human health risks from HAPs, and to meet the requirements for reducing those risks stated in the 1990 CAAA, direct measurements are needed to define the exposure of the general population to HAPs in the open atmosphere. Such "ambient air" measurements make up the second broad area of HAPs monitoring required by the CAAA. For example, the CAAA calls for a 75% reduction in the incidence of cancer caused by HAPs emitted from area sources. Knowledge of the ambient concentrations of HAPs is clearly required to estimate current health risks and to assess progress toward reducing those risks. Not surprisingly, the sampling and analytical methods applicable to ambient HAPs often differ from those used for source-related HAPs measurements.

The subject of this chapter is a compilation of existing and potential sampling and analysis methods for HAPs in ambient air. The present focus on ambient methods does not imply any value judgment regarding source-related HAPs measurements; indeed, the two areas of measurement are complementary and equally valuable. However, it must be noted that some confusion exists over the meaning of the term "ambient measurement." In this chapter, the commonly accepted definition recognized by USEPA is used: *an ambient measurement is one made in the open atmosphere, in a location removed from direct impact of an emission source, and suitable for estimating the non-occupational pollutant exposures of the general population.* On the other hand, the regulated industrial community often divides air pollution measurements into three categories: in-stack emission measurements, workplace airborne exposure measurements, and "ambient" measurements. By that definition, any pollutant determination made outdoors is denoted as an "ambient" measurement, whether in a plume, at the fence line of a facility, or in an urban neighborhood. Clearly, this definition is overly broad. It must be stressed that this chapter focuses on methods suitable for obtaining data on ambient population exposures to HAPs. Although some of the methods cited here may also be useful in other non-ambient outdoor applications, that use is not the focus of this chapter. Conversely, methods suitable for some outdoor applications may not be included here, because they are not appropriate for true ambient HAPs determination.

A primary characteristic of ambient HAPs measurements is that the levels to be measured are generally much lower than those found in or near large emission sources. The ambient HAPs concentration data presented in Chapter 4 show that HAPs levels are typically below 10 micrograms per cubic meter ($\mu g/m^3$), and often below 1 $\mu g/m^3$. (These units of concentration are readily converted to mixing ratios on a volume/volume basis, such as parts-per-billion by volume (ppbv = 1×10^{-9} v/v.) For example, at normal conditions (i.e., 20° C and one atmosphere pressure), 1 ppbv =(0.0416 • MW) $\mu g/m^3$, where MW is the molecular weight of the species.) Clearly, detection of very low concentrations is a prime requirement of any ambient method. Chapter 4 also shows that, for many of the HAPs, ambient data are nonexistent or extremely scarce. A likely cause for the scarcity of ambient HAPs data is the lack of sufficiently sensitive sampling and analysis methods for ambient measurements.

For purposes of human health risk assessment, it is also necessary that ambient measurements be conducted at sites that represent the local population and pollutant exposure distributions. In practice, this generally means that ambient measurements of HAPs are made at multiple sites within a populated area. For that type of ambient sampling network, simple, reliable, inexpensive, and broadly applicable methods are advantageous.

This chapter describes the procedures used in identifying ambient HAPs methods, presents the survey results in detail for each HAP, and also summarizes important features of the results. In addition, the following section presents some background on ambient measurement methods and puts the present information in the context of other studies.

3.2 BACKGROUND

Ambient methods development for hazardous air pollutants has been the subject of considerable research in recent years, resulting in the variety of current measurement methods available particularly for volatile organics, semivolatile organics, and particulate-phase inorganics. However, as noted in Chapter 1 and detailed in Chapter 2, the 188 HAPs are an extremely diverse group of chemicals, and include several compounds not previously considered as ambient air pollutants. Previous reviews of possible measurement methods for the 188 HAPs have generally considered only long-established standard methods, to the exclusion of novel research methods. Such reviews have generally taken optimistic views of the effectiveness of standard methods for measuring the diverse HAPs.[2,3] Furthermore, the chemical and physical properties of the individual HAPs have not been carefully considered in previous reviews. Instead, the approach generally taken was to suggest measurement methods for HAPs based on the perceived similarity of one HAP to another. The diversity of the HAPs makes such an approach suspect. The collection of information in this chapter was designed to avoid that shortcoming of previous surveys by considering HAPs properties in identifying measurement methods for the HAPs.

The diversity of the 188 HAPs is illustrated by the range of physical and chemical properties presented in Chapter 2, and by the atmospheric lifetimes and reaction pathways reported in Chapter 5. Those properties and reactivity determine the types of sampling and analysis methods suitable for each HAP in ambient air, and also allow the HAPs to be categorized for identification of generic types of sampling and analysis methods. A key factor is the vapor pressure of a HAP, which determines whether it is sufficiently volatile to be present entirely in the vapor phase in the atmosphere, or exists in both vapor and particle phases (i.e., a semivolatile compound), or in the particle phase only (a nonvolatile compound). The phase distribution in turn determines what collection media and sample storage procedures may be suitable for that compound. Other properties may then modify the primary choice of sampling approach that was based on volatility alone. For example, storage of air in a sampling canister may be unsuitable for a highly volatile compound that is also highly polar and water soluble. Similarly, reactivity with water or with other chemicals in the sample may come into play, even though the sample collection method used is appropriate for the phase distribution of the HAP in question. It is not the purpose of this chapter to review all such considerations, but extensive information is available elsewhere specifically addressing the subject of atmospheric sampling.[4,5]

It must be stressed that application of any sampling or analysis method for HAPs must consider not only the properties of the target compounds and the conditions of sampling, but also the nature of the overall sampling program, the intended use of the data, the meteorological conditions, and site characteristics. In other words, selection of sampling and analysis methods for ambient HAPs determination must be conducted as an integral part of a properly designed measurement study. The methods survey presented in this chapter can serve as a guide to appropriate sampling and analysis techniques for the HAPs, but responsibility for properly integrating and applying those techniques rests with the user. This responsibility is especially important in air monitoring programs, in contrast to research-type measurements, because "monitoring" generally implies a routine, long-term effort with potential regulatory implications as well as cost, data quality, and legal considerations.

In compiling information on methods that have been or could be used for ambient HAPs measurements, two rapidly developing approaches were noted that, at present, appear unsuitable for routine ambient monitoring activities, but that deserve special mention because of their potential advantages. Those methods are: (1) long-path optical techniques (such as Fourier-transform infrared spectroscopy (FTIR)), and (2) direct air sampling mass spectrometry (MS).

Long-path optical methods including FTIR have been used successfully for some time in source-related measurements of a number of chemicals, including some HAPs. The information gathered in this survey indicated that the detection limits and spectral databases of long-path methods are currently insufficient for detection of diverse HAPs at ambient levels. Furthermore, the complexity and costs of optical methods are generally greater than those of the sample collection techniques cited in this survey. These factors make long-path methods unattractive at present for ambient sampling networks addressing pollutant exposures of the general population. However, with further development, these methods have the potential for simultaneous determination of multiple species in nearly real time. At present, there is insufficient documentation of the ambient HAPs capabilities of long-path methods to merit inclusion of such methods in this database. However, the potential for rapid determination of multiple HAPs is a strong argument for further development of long-path methods. Support for such development is indicated, for example, by the publication of U.S.EPA's Method TO-16, addressing the use of FTIR for air pollutant measurements.[6]

Direct air sampling mass spectrometry (MS) is a much newer technology than long-path optical methods, but shows promise for rapid, highly specific, multi-component determination of HAPs in air. Direct air sampling with an atmospheric pressure chemical ionization (APCI) inlet has been implemented for HAPs and other chemicals with commercial triple quadrupole instruments.[7–9] More recently, the small size and high sensitivity of ion-trap MS instruments have led to adaptation of direct air sampling interfaces for such systems. Using a compact commercial ion trap instrument, both a polymer membrane and a glow discharge sample inlet/ionization source have been demonstrated to provide detection limits in the sub-ppbv range in continuous monitoring for some HAPs.[10–12] Furthermore, the development of software to facilitate mass isolation has made commercial ion trap instruments capable of true MS/MS analysis. Issues of cost and instrumental complexity limit the application of direct MS methods in monitoring networks, and much further development is needed. However, the specificity, sensitivity, rapid response, and potentially wide applicability of MS techniques for HAPs suggest that ambient measurements by such techniques may soon be commonplace.

3.3 SURVEY APPROACH

The survey described here differed substantially from previous reviews[2,3] of possible measurement methods for the 188 HAPs in both approach and scope. A highlight of the current approach was the initial compilation of key physical and chemical properties of the HAPs, as presented in Chapter 2. These properties were used to group the HAPs into various classes of compounds and, subsequently, to conduct evaluations of the applicability of individual measurement methods.

The search for measurement methods for the HAPs was intended to be as wide ranging as possible. Information sources included standard compilations of air sampling methods, such as EPA Screening Methods, EPA Contract Laboratory Program (CLP) and Compendium methods, as have been used in previous surveys.[13-15] However, this study also reviewed standard methods designated by the Intersociety Committee on Methods of Air Sampling and Analysis, the National Institute of Occupational Safety and Health (NIOSH), the Occupational Safety and Health Administration (OSHA), the American Society for Testing and Materials (ASTM), and the EPA Compendium IO-Methods. Although not necessarily targeted for ambient air measurements, these methods are well documented and might serve as the starting point for an ambient air method. EPA solid waste (SW 846) methods were also consulted. Another resource was the EPA database on measurement methods for HAPs,[13] which primarily includes established EPA methods. Additional sources of information were surveys on the ambient concentrations[16] and atmospheric transformations[16-18] of the HAPs. Those surveys are presented as Chapters 4 and 5 of this book. The ambient concentrations surveys[16,19-21] were especially useful as a guide to measurement methods for HAPs, and assured that methods were identified for all HAPs that have been measured in ambient air. In addition, reports, journal articles, and meeting proceedings known to contain information on HAPs methods were obtained and reviewed.

A unique feature of this survey was the evaluation of the state of development of individual HAPs measurement methods, distinguishing workplace, laboratory or stack emission methods from methods actually tested in ambient air. The extent of documentation and actual ambient use of methods were key considerations in making that distinction. The measurement methods identified for the 188 HAPs were organized into three categories: *applicable*, *likely*, and *potential*.[13,14]

Applicable — An applicable method was defined as one that has been reasonably established and documented for measurement of the target HAP in ambient air. In most cases, methods identified as applicable have actually been used for ambient measurements, i.e., ambient data are available illustrating the effectiveness of the method. A good example of an applicable method is EPA Compendium Method TO-14A, which has been widely used for VOC measurements.[22] In other cases, a method was identified as applicable for a specific HAP because of the degree of documentation and standardization of the method, even though no ambient data were found. The primary examples of this are a few CLP and TO- methods. Although such methods are targeted for a number of HAPs, for a few of those HAPs no ambient measurements were found, and further development may be needed to achieve ambient measurement capabilities. It must be stressed that the existence of an applicable method does not guarantee adequate measurement of the pertinent HAP(s) under all circumstances. Further development and evaluation may be needed to assure sensitivity, freedom from interferences, stability of samples, precision, accuracy, etc. under the range of conditions found in ambient measurements.

Likely — Two types of likely measurement methods were defined. The most common is a method that has been clearly established and used for the target HAP in air, but not in ambient air. The presumption is that further development (such as an increase in sensitivity or sampled volume) would allow measurements in ambient air. The primary examples of this type of likely method are NIOSH or OSHA methods established for HAPs in workplace air. A specific example is OSHA Method No. 21, stated to have a detection limit of 1.3 ppbv in workplace air, and designated as a likely method for acrylamide. In a few cases, such methods have been applied to ambient air, but in such limited conditions or time periods, that demonstration of the method is judged to be incomplete. The second type of likely method consists of techniques identified as applicable for one HAP, and consequently inferred as likely for another, based on close similarity of chemical and physical properties. An example of an inferred likely method is TO-14A for 1,2-dibromo-3-

chloropropane, based on the similarity of this compound to other VOCs in terms of volatility, water solubility, and reactivity.

Potential — A potential method was defined as one that needs extensive further development before application to ambient air measurements is justified. Many potential methods have been evaluated under laboratory conditions, or for the target HAP in sample matrices other than air (e.g., water, soil). Potential methods were inferred for some HAPs, based on applicable or likely methods found for other HAPs of somewhat similar chemical and physical properties. The degree of similarity of properties between HAPs was used as the guide in designating potential methods in those cases.

For HAPs for which no applicable or likely methods were found, further searches were conducted beyond the reviews outlined above. For such HAPs, detailed literature searches were conducted using the computer database files of Chemical Abstracts Service (CAS) and the National Technical Information Service (NTIS). Methods identified through such searches were then subjected to the same evaluation and categorization standards.

In all method searches and reviews, the chemical and physical properties compiled in Chapter 2 were valuable. The quantitative similarity of properties such as vapor pressure, solubility, and reactivity of HAPs was used to suggest likely and potential methods, and the degree of similarity of properties determined the choice between designation as a likely or potential method. In compiling information on measurement methods, HAPs consisting of compound classes (e.g., PCBs, PAHs) were addressed by identifying methods for the most and least volatile species of each class likely to be present in ambient air. For each HAP, all identified methods are categorized as applicable, likely, or potential methods, and listed using standard method designations (e.g., TO-5, CLP-2, NIOSH 5514), or by citations of the pertinent literature (e.g., R-1, R-2).

A key characteristic of an ambient air measurement method is the detection limit. As part of this methods survey, ambient air detection limits were indicated whenever they were reported in method documentation. The various units in which detection limits were reported include mixing ratios in parts-per-million by volume (ppmv), parts-per-billion by volume (ppbv), and parts-per-trillion by volume (pptv), and mass concentrations in milligrams per cubic meter (mg/m^3), micrograms per cubic meter ($\mu g/m^3$), nanograms per cubic meter (ng/m^3), and picograms per cubic meter (pg/m^3). The means of interconverting between these two sets of units is given in section 3.1. Detection limits were reported in this review as they were stated in the respective methods. Detection limits for certain CLP methods were reported as contract required quantitation limits (CRQL) in mass units only (e.g., ng), or as a range of applicable concentrations. In such cases, the detection limit was reported as stated in the method, along with needed supporting information such as the approximate sampled air volume. An effort was made to indicate the detection limit for at least the most fully developed method(s) for each HAP. Estimation of detection limits, when they were not explicitly stated in the material reviewed, was generally not done. The detection limits reported should be considered primarily as guides to the relative capabilities of the various methods, rather than as absolute statements of method performance.

Citation of literature was aimed at providing the user enough information to review at least the basics of the identified method, and to locate further information if needed. No effort was made to compile all possible information on each method.

3.4 STATUS OF CURRENT METHODS

This survey identified more than 300 methods pertinent to ambient measurements of the 188 HAPs, comprising TO- methods, IO- methods, NIOSH methods, OSHA methods, EPA screening methods, CLP methods, and research methods published in the open literature. The complete results of the HAPs method survey are presented in Table 3.1 (see Appendix following Chapter 3), which lists the 188 HAPs in the same order as they appear in the CAAA, and, for each HAP, shows the CAS number, the volatility class, and indications of the pertinent ambient methods. The ambient methods information is listed in successive columns for applicable, likely, and potential methods. Within each of these columns, the identified methods are indicated by standard method designations (e.g., TO-5, CLP-2, OSHA CIM [0065]), or by citations of the pertinent research literature (e.g., R-1, R-2). The final two columns of the table show the detection limits for selected methods, and provide explanatory comments on the entries, respectively.

A list of all the methods and literature cited in Table 3.1 is appended. Standard methods, such as NIOSH, OSHA, or TO- methods, are listed by title under a general reference heading. Research methods are listed in numerical order (R-1, R-2, etc.). For each research method, the citation includes a brief description of the method and one or more literature citations pertinent to the method. The reader is referred to Table 3.1 for the full results of the methods survey. However, some general comments on the findings of this study are of interest here.

Figure 3.1 shows that, for 134 HAPs (two thirds of the HAPs list), applicable ambient measurement methods were found. Note that it shows only the most developed state of methods found; for some of these 134 HAPs, likely and potential methods were also found. Figure 3.1 also shows that, for 43 HAPs, likely methods were found, but no applicable methods. Most of these likely methods were specific for the HAP in question, but for some, the identification of likely methods was inferred based on HAP properties. For nine HAPS only potential methods could be identified, and of those, three were inferred on the basis of chemical and physical properties. For two HAPs (ethyl carbamate and titanium tetrachloride), no measurement methods could be identified at any level of development.

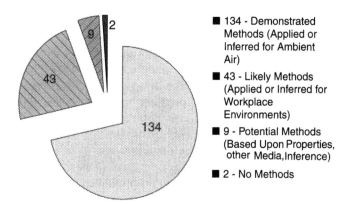

FIGURE 3.1 Distribution of the 188 HAPs by the most developed type of ambient measurement method currently available for each compound.

3.5 HAPS METHOD DEVELOPMENT: FUTURE DIRECTIONS

In terms of method development needs for the HAPs, the most cost-effective approach would probably be further development of the likely methods that exist for the HAPs with no applicable methods. The definition of a likely method means that a reasonable degree of further development should result in a method applicable to ambient air. In addition, the large number of applicable methods already available for volatile and semivolatile organics should enhance development of methods for additional compounds. A good example is the TO-15 document, which discusses canister sampling and its potential for sampling the 97 volatile HAPs.[23] Validation on storage stability and analytical method detection needs to be determined for many of these compounds.[24–29]

Continued evaluation of measurement methods for all the HAPs would be worthwhile. An important goal of that effort should be to consolidate and simplify the variety of methods available into a smaller number of well-characterized and broadly applicable methods. Although some of the standard EPA methods cited in this survey are intended to be broadly applicable, the diversity

3.6 SUMMARY

This chapter presents the status of ambient air measurement methods for the 188 HAPs. Over 300 different candidate measurement methods currently in various stages of development are cited. Only 134 of the 188 HAPs have methods that are reasonably established for ambient air measurements. However, even these reasonably established methods are not necessarily all EPA-approved or fully demonstrated for ambient monitoring. Of the remaining HAPs, 43 have methods that are reasonably established for non-ambient air, such as for workplace or stack emission measurements, and could likely be developed for ambient air applications. Of the 11 remaining HAPs, nine have methods that could potentially be applicable to ambient air measurements following extensive further development, and two have no methods currently in any stage of development. These findings point to the need for continued methods development to address the measurement gaps identified, and to consolidate the many similar methods found into more broadly capable methods.

REFERENCES

1. Clean Air Act Amendments of 1990, Conference Report to Accompany S. 1630, Report No. 101-952, U.S. Government Printing Office, Washington, D.C., 139, 1990.
2. Keith, L.H. and Walker, M.M., *EPA's Clean Air Act Air Toxics Database, Volume I: Sampling and Analysis Methods Summaries,* Lewis, Boca Raton, FL, 1992.
3. Winberry, W.T., Jr., Sampling and analysis under Title III, *Environmental Lab.*, 46, June/July 1988.
4. Ambient air sampling information available at www.epa.gov/ttn/amtic.
5. Coutant, R.W. and McClenny, W.A., Competitive adsorption effects and the stability of VOC and PVOC in canisters, in *Proc. 1991 EPA/AWMA Symp. Measurement of Toxic and Related Air Pollutants,* EPA-600/9-91/018, Publication No. VIP-21, Air and Waste Management Association, Pittsburgh, PA, 382, 1991.
6. TO-16 document available at www.epa.gov/ttn/amtic/airtox.html.
7. Dawson, P.H. et al., The use of triple quadrupoles for sequential mass spectrometry 1: the instrument parameters, *Org. Mass Spectrom.*, 17, 205, 1982.
8. Busch, K.L., Glish, G.L. and McLuckey, S.A., *Mass Spectrometry/Mass Spectrometry: Techniques and Applications of Tandem Mass Spectrometry,* John Wiley & Sons., New York, 1989.
9. Kelly, T.J. and Kenny, D.V., Continuous determination of dimethylsulfide at part-per-trillion concentrations in air by atmospheric pressure chemical ionization mass spectrometry, *Atmos. Environ.*, 10, 2155, 1991.
10. Wise, M.B. et al., Review of direct MS analysis of environmental samples, *Field Anal. Chem. Technol.*, 1, 251, 1997.
11. McLuckey, S.A., Glish, G.L., and Asano, K.G., The coupling of an atmospheric sampling ion source with an ion trap mass spectrometer, *Anal. Chim. Acta*, 225, 25, 1989.
12. Gordon, S.M. et al., Direct sampling and analysis of volatile organic compounds in air by membrane introduction and glow discharge ion trap mass spectrometry with filtered noise fields, *Rapid Commun. Mass Spectrom.*, 10, 1038, 1996.
13. Kelly, T.J. et al., Ambient measurement methods and properties of the 189 Clean Air Act hazardous air pollutants, Final Report to U.S. EPA, EPA-600R-94/187, Battelle, Columbus, OH, March 1994.
14. Holdren, M.W., Abbgy, S. and Armbruster, M.J., Ambient measurement methods and properties of the 188 Clean Air Act hazardous air pollutants, Final Report to U.S. EPA, Contract 68-D-98-030, Work Assignment No. 1-Task 4, Battelle, Columbus, OH, March 1999.
15. Mukund, R. et al., Status of ambient measurement methods for hazardous air pollutants, *Environ. Sci. Technol.*, 29, 183A-187A, 1995.
16. Kelly, T.J. et al., Concentrations and transformations of hazardous air pollutants, *Environ. Sci. Technol.*, 28, 378A, 1994.
17. Spicer, C.W. et al., A literature review of atmospheric transformation products of Clean Air Act Title III Hazardous Air Pollutants, Final Report to U.S. EPA, EPA-600/R-94-088, Battelle, Columbus, Ohio, July 1993.

18. Kelly, T.J. et al., Surveys of the 189 CAAA Hazardous Air Pollutants: II. Atmospheric Lifetimes and Transformation Products, in *Measurement of Toxic and Related Air Pollutants, Proc. 1993 EPA/AWMA Int. Symp.*, EPA Report No. EPA/600/A93/024, Publication VIP-34, Air and Waste Management Association, Pittsburgh, PA, 167, 1993.
19. Shah, J.J. and Heyerdahl, E.K., National ambient volatile organic compounds (VOCs) database update, Report EPA-600/3-88/01(a), U.S. EPA, Research Triangle Park, NC, 1988.
20. Shah, J.J. and Singh, H.B, Distribution of volatile organic chemicals in outdoor and indoor air: A national VOCs database, *Environ. Sci. Technol.*, 22, 1381, 1988.
21. Shah, J.J. and Joseph, D.W. National ambient VOC data base update: 3.0, report to U.S. EPA, EPA-600/R-94-089, by G_2 Environmental, Inc., Washington, D.C., under subcontract from Battelle, Columbus, OH, May 1993.
22. McClenny, W.A. et al., Canister-based method for monitoring toxic VOCs in ambient air, *J. Air Waste Manage. Assoc.*, 41, 1308, 1991.
23. TO-15 document available at www.epa.gov/ttn/amtic/airtox.html.
24. McClenny, W.A. et al., Status of VOC methods development to meet monitoring requirements for the Clean Air Act Amendments of 1990, in *Measurement of Toxic and Related Air Pollutants, Proc. 1991 EPA/AWMA Int. Symp.*,, Report No. EPA-600/9-91/018, Publication VIP-21, *Air and Waste Management Assoc.*, Pittsburgh, PA, 367, 1991.
25. Kelly, T.J. and Holdren, M.W., Applicability of canisters for sample storage in the determination of hazardous air pollutants, *Atmos. Environ.*, 29, 2595, 1995.
26. Kelly, T.J. et al., Method development and field measurements for polar volatile organic compounds in ambient air, *Environ. Sci. Technol.*, 27, 1146, 1993.
27. Oliver, K.D. Sample integrity of trace level polar VOCs in ambient air stored in summa-polished canisters, Technical Note TN-4420-93-03, submitted to U.S. EPA under Contract No. 68-D0-0106, by ManTech Environmental Technology, Inc., Research Triangle Park, NC, Nov., 1993.
28. Pate, B. et al., Temporal stability of polar organic compounds in stainless steel canisters, *J. Air Waste Manage. Assoc.*, 42, 460, 1992.
29. Coutant, R.W., Theoretical evaluation of stability of volatile organic chemicals and polar volatile organic chemicals in canisters, Final Report to U.S. EPA, Contract No. 68-D0-0007, Work Assignment No. 45, Subtask 2, Battelle, Columbus, OH, September 1993.

APPENDIX

TABLE A3.1
Results of the Survey of Ambient Air Measurement Methods for the 188 HAPs (Chemicals shown in italics are high priority urban HAPS)

Compound	CAS No.	Compound Class[a]	Ambient Measurement Method			Limit of Detection	Comment
			Applicable	Likely	Potential		
Acetaldehyde	75-07-0	VVOC	TO-11A	R-4 [14] OSHA 68 NIOSH 2538 NIOSH 2539 NIOSH 3507		TO-11A: 1 ppbv [14]: 30 ppmv [2538]: 2 µg/sample [3507]: 0.1 mg/sample [68]: 580 ppb (1050 µg/m³)	
Acetamide	60-35-5	SVOC			OSHA A625 R-37 R-47		R-47: method developed for analysis of water
Acetonitrile	75-05-8	VOC	TO-15 TO-17 R-1 R-3	NIOSH 1606		TO-17: ≤ 0.5 ppb R-1: 1 ppbv [1606]: 0.8 µg/sample	
Acetophenone	98-86-2	VOC			OSHA A169		
2-Acetylaminofluorene	53-96-3	NVOC			OSHA 0065		
Acrolein	107-02-8	VOC	TO-11A	OSHA 52 NIOSH 2501 NIOSH 2539		TO-11A: 1 ppbv [2501]: 2 µg/sample [52]: 2.7 ppb (6.1 µg/m³)	
Acrylamide	79-06-1	VOC		OSHA 21		[21]: 1.3 ppbv	
Acrylic acid	79-10-7	VOC		OSHA 28		[28]: 42 µg/m³ (14 ppbv)	

Compound	CAS	Class	Methods	Methods		Detection limits	Notes
Acrylonitrile	107-13-1	VOC	TO-15 TO-17 R-1 R-3	OSHA 37 NIOSH 1604 R-4 [14]		R-1: 1 ppbv TO-17: ≤0.5 ppbv [1604]: 1 µg/sample [37]: 0.026 mg/m³ (0.1 ppm)	
Allyl chloride	107-05-1	VOC	TO-14A TO-15 R-3	NIOSH 1000		TO-14A: 0.1 ppbv 0.01 mg/sample	
4-Aminobiphenyl	92-67-1	SVOC		OSHA 93 R-36	R-37	R-36: 0.1 ng/m³ [93]: 1 ppt (6.9 ng/m³)	R-36: evaluated for particulate phase only
Aniline	62-53-3	VOC	TO-15 TO-17	NIOSH 2002 NIOSH 2017	OSHA 0220	TO-17: ≤0.5 ppb [2002]: 0.01 mg/sample	
o-Anisidine	90-04-0	SVOC		NIOSH 2514	OSHA 0225	[2514]: 0.35 µg/sample [0225]: 0.06 mg/m³	[2514]: working range = 0.06–0.8 mg/m³ (200L sample volume)
Asbestos	1332-21-4	NVINC	R-21	NIOSH 7400 NIOSH 7402 NIOSH 9000 NIOSH 9002 OSHA ID160 R-63		R-21: < 0.1 ng/m³ (i.e., < 0.01 fibers/cc) [7400]: 7 fibers/mm² filter area [9002]: < 1% asbestos [ID160]: 5.5 fibers/mm²	[7400] & [7402]: working range = 0.04–0.5 fiber/cc (1000-L sample volume)
Benzene	71-43-2	VOC	TO-14A TO-15 TO-17 R-1 R-3 R-6	OSHA 12 NIOSH 1500 NIOSH 1501 NIOSH 3700 NIOSH 2549		TO-14A: 0.1 ppbv TO-17: ≤0.5 ppb [1500]: 0.001 to 0.01 mg/sample with capillary column [3700]: 0.01 ppm for 1-ml injection [1501]: 0.001 to 0.01 mg/sample with capillary column	

TABLE A3.1
Results of the Survey of Ambient Air Measurement Methods for the 188 HAPs (Chemicals shown in italics are high priority urban HAPS)

Compound	CAS No.	Compound Class[a]	Ambient Measurement Method			Limit of Detection	Comment
			Applicable	Likely	Potential		
Benzidine	92-87-5	SVOC		OSHA 65, NIOSH 5509, R-36	R-37	R-36: 1 ng/m³; [5509]: 0.05 µg/sample; [65]: 31 ng/m³	R-36: Evaluated for particulate phase only
Benzotrichloride	98-07-7	SVOC			OSHA B408		
Benzyl chloride	100-44-7	VOC	TO-14A, TO-15, R-3	NIOSH 1003		TO-14A: 0.1 ppbv; [1003]: 0.01 mg/sample	
Biphenyl	92-52-4	SVOC	R-50, R-51	NIOSH 2530	OSHA 1011	R-50: 14–16 ng/m³; [2530]: 0.09 µg/sample	[2530]: working range = 0.13–4 mg/m³ (30-L sample volume); R-50: LOD is range of ambient data
Bis(2-ethyl hexyl)phthalate (DEHP)	117-81-7	SVOC	R-28, R-57		OSHA 1015	R-28: 0.77–3.60 ng/m³	R-28: LOD shown is range of reported ambient data
Bis(chloromethyl) ether	542-88-1	VOC	TO-15	OSHA 10		[10]: 1 µg/m³	
Bromoform	75-25-2	VOC	TO-15	NIOSH 1003	OSHA 0400	[1003]: 0.01 mg/sample	
1,3-Butadiene	*106-99-0*	*VVOC*	*TO-15, R-1, R-3*	*OSHA 56, NIOSH 1024, TO-14A*		*R-1: 1 ppbv; [1024]: 0.2 µg/sample; [56]: 90 ppb (200 µg/m³)*	*[1024]: working range = 0.02–8.4 ppmv (25-L sample volume)*

Name	CAS	Class	Method	Method	OSHA	Values	Notes
Calcium cyanamide	156-62-7	Particulate		R-32	OSHA 0510	R-32: 0.08 mg/m³	R-32: recommended range in air = 0.24 mg/m³ (240-L sample volume)
Captan	133-06-2	SVOC	TO-10A R-27		OSHA 0529	TO-10A: 0.01–50 µg/m³ R-27: 1.6–14 ng/m³ [0529]: 6 ng/injection	
Carbaryl	63-25-2	SVOC	R-27	OSHA 63 NIOSH 5006		[63]: 0.028 mg/m³ R-27: 8–42 ng/m³ [5006]: 0.03 mg/sample	[5006]: working range = 0.5–20 mg/m³ (200-L sample volume) [63]: sample volume = 60 L
Carbon disulfide	75-15-0	VOC	TO-15 R-11	NIOSH 1600 R-4 [14]		R-11: 0.02 ppbv [14]: 20 ppmv [1600]: 0.02 mg/sample	LOD of R-11 estimated; range of ambient data 0.025–0.34 ppbv
Carbon tetrachloride	56-23-5	VOC	TO-14A TO-15 TO-17 R-3 R-6	NIOSH 1003		TO-14A: 0.1 ppbv TO-17: ≤0.5 ppb [1003]: 0.01 mg/sample	
Carbonyl sulfide	463-58-1	VVOC	R-10	R-4 [14]	OSHA R220	R-10: 0.03 ppbv [14]: 1 ppmv	LOD for R-10 estimated based on calibration data; range of ambient data 0.4–0.7 ppbv
Catechol	120-80-9	VOC	TO-8	R-2 R-25	OSHA 0571	TO-8: 1-5 ppbv	R-2 indicated by analogy with phenol based on similar properties R-2: 0.02 ppbv (estimated) R-25: 1 ppbv (estimated)
Chloramben	133-90-4	SVOC		R-27			R-27 indicated based on applicability of method for other pesticides

TABLE A3.1
Results of the Survey of Ambient Air Measurement Methods for the 188 HAPs (Chemicals shown in italics are high priority urban HAPS)

Compound	CAS No.	Compound Class[a]	Ambient Measurement Method			Limit of Detection	Comment
			Applicable	Likely	Potential		
Chlordane	57-74-9	SVOC	TO-10A R-27 R-28 R-29 R-30 R-31	OSHA 67 NIOSH 5510		TO-10A: 0.01-50 µg/m³ R-27: 4–50 ng/m³ R-29: < 5 pg/m³ [5510]: 0.1 µg/sample [67]: 0.064 µg/m³	
Chlorine	7782-50-5	VVINC	R-4 [805]	OSHA ID101 NIOSH 6011		[ID101]: 14 ppbv	[805]: LOD not established [6011]: working range = 7–500 ppbv (90-L sample volume) [ID101]: sample volume = 15 L
Chloroacetic acid	79-11-8	VOC		R-42 NIOSH 2008		R-42: 0.2 mg/m³ (51 ppbv) [2008]: 0.04 µg/sample	
2-Chloroacetophenone	532-27-4	SVOC		NIOSH II [291]	OSHA 0618	[291]: 0.18–0.62 mg/m³	[291]: sample volume = 12 L (measurement range shown as LOD)
Chlorobenzene	108-90-7	VOC	TO-14A TO-15 TO-17 R-3	NIOSH 1003		TO-14A: 0.1 ppbv TO-17: ≤ 0.5 ppb [1003]: 0.01 mg/sample	

Name	CAS	Class	Method	Method	OSHA	LOD	Notes
Chlorobenzilate	510-15-6	SVOC	TO-10A	R-46	OSHA 1113 R-27	TO-10A: 0.01-50 µg/m³	R-46: No LODs or air concentrations reported (workplace exposure measurements)
Chloroform	*67-66-3*	*VOC*	*TO-14A* *TO-15 R-6*	*OSHA 5* *NIOSH 1003*		*TO-14A: 0.1 ppbv* *[1003]: 0.01 mg/sample* *[5]: 0.11 ppm*	
Chloromethyl methyl ether	107-30-2	VOC		OSHA 10 NIOSH 220 R-56		[220]: 0.5 ppbv R-56: 1 ppbv [10]: 0.8 µg/m³	[220]: sample volume = 10 L (measurement range shown as LOD)
Chloroprene	126-99-8	VOC	TO-15 R-7	OSHA 112 NIOSH 1002		R-7: 0.06 ppbv [1002]: 0.03 mg/sample [112]: 22 ppb (80 µg/m³)	
Cresol/Cresylic acid (mixed isomers)	1319-77-3	VOC	TO-8	OSHA 32 NIOSH 2549 NIOSH 2546 R-60	OSHA 0760	TO-8: 1-5 ppbv [2546]: 1 to 3 µg/sample [32]: 0.046 mg/m³ (0.01 ppm) [0760]: 14 ng/sample	
o-Cresol	95-48-7	VOC	TO-8	OSHA 32 NIOSH 2549 NIOSH 2546 R-2 R-25 R-60	OSHA 0760 R-59	TO-8: 1-5 ppbv [2546]: 1 to 3 µg/sample [32]: 0.046 mg/m³ (0.01 ppm) [0760]: 14 ng/sample	R-2: 0.02 ppbv (estimated)
m-Cresol	108-39-4	SVOC	TO-8	OSHA 32 NIOSH 2549 NIOSH 2546 R-2 R-3 R-60	OSHA 0760 R-59	TO-8: 4.5-22.5 µg/m³ R-2: 4.5 µg/m³ R-3: 0.09 µg/m³ [2546]: 1 to 3 µg/sample [32]: 0.046 mg/m³ (0.01 ppm) [0760]: 14 ng/sample	

TABLE A3.1
Results of the Survey of Ambient Air Measurement Methods for the 188 HAPs (Chemicals shown in italics are high priority urban HAPS)

Compound	CAS No.	Compound Class[a]	Ambient Measurement Method			Limit of Detection	Comment
			Applicable	Likely	Potential		
p-Cresol	106-44-5	SVOC	TO-8	OSHA 32 NIOSH 2549 NIOSH 2546 R-2 R-3 R-59 R-60	OSHA 0760	TO-8: 4.5–22.5 µg/m³ R-2: 4.5 µg/m³ R-3: 0.09 µg/m³ [2546]: 1 to 3 µg/sample [32]: 0.046 mg/m³ (0.01 ppm) [0760]: 14 ng/sample	
Cumene	98-82-8	VOC	TO-15 TO-14A R-6	NIOSH 1501		TO-14A: 0.1 ppbv [1501]: 0.001 to 0.01 mg/sample with capillary column	
2,4-D (2,4-Dichloro phenoxyacetic acid) (incl. salts and esters)	N/A	SVOC	TO-10A R-27	NIOSH 5001	R-38	TO-10A: 0.01–50 µg/m³ R-27: < 0.8 ng/m³ [5001]: 0.015 mg/filter	T-10A only for esters; 2,4-acid and salts would require filter for particulate; see R-38 [5001]: working range = 1.5–20 mg/m³ (100-L sample volume) R-27: esters only
DDE (1,1-dichloro-2,2-bis(p-chloro phenyl)ethylene)	72-55-9	SVOC	TO-10A R-29 R-27 R-28			TO-10: 0.01- 50 ng/m³ R-29: < 5 pg/m³ R-27: 1.4–3.6 ng/m³	

Compound	CAS	Class	Methods		OSHA	LOD	Notes
Diazomethane	334-88-3	VVOC		NIOSH 2515	OSHA 0861	[2515]: LOD not determined	[2515]: working range = 0.11–0.57 ppmv (10-L sample volume)
Dibenzofuran	132-64-9	SVOC	TO-9A R-50 R-5 R-51	R-4 [836]	OSHA D639	TO-9A: 1-5 pg/m^3 [836]: 3.3 ng/m^3 R-50: 13–26 ng/m^3 R-5: 0.02 pg/m^3 R-51: < 0.01 pg/m^3 [D639]: 2.3 ng/injection	[836]: sample volume = 1500 m^3, method is for total particulate aromatic hydrocarbons R-50: LOD is range of ambient data. Higher chlorinated species (e.g., octa-) are probably NVOC
1,2-Dibromo-3-chloropropane	96-12-8	VOC	TO-15 R-12	TO-14A	OSHA 0935	R-12: < 2 ng/m^3 (< 0.2 pptv) TO-14A: 0.1 ppbv	TO-14A, TO-15 indicated by analogy with VOCs having similar properties R-12: range of ambient data 2–21 ng/m^3
Dibutyl phthalate	84-74-2	SVOC	R-28 R-57	OSHA 104 NIOSH 5020		R-28: 0.48–3.6 ng/m^3 R-57: 5–370 ng/m^3 [5020]: 10 µg/sample [104]: 34 µg/m^3	R-28: LOD shown is range of reported ambient data R-57: LOD shown is range of ambient data for separate vapor and particulate measurements of various isomers
1,4-Dichlorobenzene	106-46-7	VOC	TO-14A TO-15 R-3	NIOSH 1003		TO-14A: 0.1 ppbv [1003]: 0.01 mg/sample	

TABLE A3.1
Results of the Survey of Ambient Air Measurement Methods for the 188 HAPs (Chemicals shown in italics are high priority urban HAPS)

Compound	CAS No.	Compound Class[a]	Ambient Measurement Method			Limit of Detection	Comment
			Applicable	Likely	Potential		
3,3'-Dichlorobenzidine	91-94-1	SVOC		NIOSH 5509 OSHA 65 R-36	R-37	[65]: 40 ng/m³ R-36: 0.1 ng/m³ [5509]: 0.05 µg/sample	[5509]: working range = 4–200 µg/m³ (50-L sample volume) [65]: sample volume = 100 L R-36: evaluated for particulate phase only
Dichloroethyl ether (Bis[2-chloroethyl]ether)	111-44-4	VOC	TO-15	NIOSH 1004		[1004]: 0.01 mg/sample	
1,3-Dichloropropene	*542-75-6*	*VOC*	*TO-15 TO-14A R-3*		*OSHA D177*	*TO-14A: 0.1 ppbv*	
Dichlorvos	62-73-7	SVOC	TO-10A R-27	OSHA 62	OSHA 0850	TO-10A: 0.01–50 µg/m³ [62]: 1.9 µg/m³ (0.21 ppb)	
Diethanolamine	111-42-2	SVOC		NIOSH 3509	OSHA D129	[3509]: 7 to 20 µg/sample [D129]: 1.6 µg/sample	[3509]: working range = 0.4–3 mg/m³ (100-L sample volume)
Diethyl sulfate	64-67-5	VOC	TO-15	R-40 R-8	OSHA 0913	R-40: 8 pptv	R-40: not applied to ambient air analysis Indication of R-8 based on similarity of properties with dimethyl sulfate

Compound	CAS	Type	Method	Methods	OSHA	Values	Comments
3,3'-Dimethoxybenzidine	119-90-4	NVOC		R-36	OSHA 0873 R-37	R-36: 1 ng/m^3 [0873]: 5 ng/injection	R-36: Evaluated for particulate phase only
4-Dimethylaminoazo-benzene	60-11-7	NVOC		NIOSH 284	OSHA 0929	[284]: 4–2000 µg/m^3 [0929]: 6 ng/injection	[284]: sample volume = 50 L (measurement range shown as LOD)
N,N-Dimethylaniline	121-69-7	VOC		NIOSH 2002	OSHA 0931	[2002]: 0.01 mg/sample	[2002]: measurement range = 0.05–3.0 mg/sample (unknown sample volume)
3,3'-Dimethylbenzidine	119-93-7	SVOC		R-36	OSHA 2450 R-37	R-36: 1 ng/m^3 [2450]: 0.01 µg/sample	R-36: evaluated for particulate phase only
Dimethylcarbamoyl chloride	79-44-7	VOC		R-39		R-39: 0.05 ppbv	R-39: sample volume = 48 L
N,N-Dimethylformamide	68-12-2	VOC	R-9	OSHA 66 NIOSH 2004 R-4 [14]		R-9: 0.6–50 ppbv [2004]: 0.05 mg/sample [66]: 0.02 ppm (0.045 mg/m^3)	R-9: reports four separate methods
1,1-Dimethylhydrazine	57-14-7	VOC	TO-15	NIOSH 3515 R-22	OSHA 0940	R-22: 4 ppbv [3515]: 1 µg/sample	[S143]: working range = 0.04–4 ppmv (100-L sample volume) R-22: sample volume = 2 L
Dimethyl phthalate	131-11-3	SVOC		OSHA 104 R-26 R-28		R-26: 60 ng/m^3 [104]: 90 fg/m^3	R-28 suggested by analogy with di-n-butyl phthalate
Dimethyl sulfate	77-78-1	VOC	TO-15 R-8	NIOSH 2524	OSHA 0960	R-8: 0.05 ppbv [2524]: 0.25 µg/sample	LOD for R-8 estimated based on ranges of sampling durations, sampling rates, and analytical capabilities

TABLE A3.1
Results of the Survey of Ambient Air Measurement Methods for the 188 HAPs (Chemicals shown in italics are high priority urban HAPS)

Compound	CAS No.	Compound Class[a]	Ambient Measurement Method			Limit of Detection	Comment
			Applicable	Likely	Potential		
4,6-Dinitro-o-cresol (including salts)	N/A	SVOC	TO-8		OSHA 0975 R-3	TO-8: 1-5 ppbv	R-3 suggested by analogy to other phenols
2,4-Dinitrophenol	51-28-5	SVOC	TO-8			TO-8: 1-5 ppbv	
2,4-Dinitrotoluene	121-14-2	SVOC		OSHA 44	OSHA 0990	[44]: 20 µg/m³	
1,4-Dioxane (1,4-Diethyleneoxide)	123-91-1	VOC	TO-15	NIOSH 1602 R-4 [14]	OSHA 1010	[14]: 2 ppmv [1602]: 0.01 mg/sample	
1,2-Diphenylhydrazine	122-66-7	SVOC			R-22		Suggestion of R-22 based on chemical similarity to volatile hydrazines
Epichlorohydrin (1-Chloro-2,3-epoxypropane)	106-89-8	VOC		NIOSH 1010 R-4 [14]	OSHA 0645	[14]: 20 ppmv [1010]: 1.0 µg/sample	
1,2-Epoxybutane	106-88-7	VOC		NIOSH 1614 R-3	OSHA E225		R-3 and NIOSH methods indicated by similarity of properties with ethylene oxide
Ethyl acrylate	140-88-5	VOC	TO-15 TO-17 R-1 R-3	OSHA 92 NIOSH 1450		R-1: 0.2 ppbv TO-17: ≤ 0.5 ppb [1450]: 0.02 mg/sample [92]: 80 µg/m³	

Compound	CAS	Class	Methods	NIOSH/OSHA	OSHA	Detection limits/notes
Ethylbenzene	100-41-4	VOC	TO-14A, TO-15, TO-17, R-3, R-6	NIOSH 1501		TO-14A: 0.1 ppbv; TO-17: ≤0.5 ppb; [1501]: 0.001 to 0.01 mg/sample with capillary column
Ethyl carbamate (urethane)	51-79-6	VOC				
Ethyl chloride	75-00-3	VVOC	TO-14A, TO-15, R-3	NIOSH 2519, R-4 [14]	OSHA 1110	TO-14A: 0.1 ppbv; [14]: 10 ppmv; [2519]: 0.01 mg/sample
Ethylene dibromide	106-93-4	VOC	TO-14A, TO-15	OSHA 2, NIOSH 1008		TO-14A: 0.1 ppbv; [1008]: 0.01 µg/sample; [2]: 0.005 mg/m³
Ethylene dichloride	107-06-2	VOC	TO-14A, TO-15, R-3	OSHA 3, NIOSH 1003		TO-14A: 0.1 ppbv; [1003]: 0.01 mg/sample; [3]: 0.05 ppm
Ethylene glycol	107-21-1	SVOC		NIOSH 5500, NIOSH 5523	OSHA 1911	[5523]: 7 µg/sample; [5500]: working range = 7–330 mg/m³ (3-L sample volume)
Ethyleneimine	151-56-4	VOC		NIOSH 3514, R-4 [14]		[14]: 15 ppmv; [3514]: 0.3 µg/sample
Ethylene oxide	75-21-8	VVOC	TO-15, R-13	OSHA 30, OSHA 49, OSHA 50, NIOSH 1614, NIOSH 3702, R-3		R-13: 0.001–0.1 ppbv; [1614]: 1 µg/sample; [3702]: 2.5 pg per 1-ml injection; [30]: 24.0 µg/m³ (13.3 ppb); [49]: 1.3 µg/m³; [1614]: working range = 0.04–4.5 ppmv (24-L sample volume); R-13 evaluated five different methods

TABLE A3.1
Results of the Survey of Ambient Air Measurement Methods for the 188 HAPs (Chemicals shown in italics are high priority urban HAPS)

Compound	CAS No.	Compound Class[a]	Ambient Measurement Method			Limit of Detection	Comment
			Applicable	Likely	Potential		
Ethylene thiourea	96-45-7	SVOC		OSHA 95 NIOSH 5011		[5011]: 0.75 µg/sample [95]: 1.39 fg/m³	[5011]: working range = 0.05–75 mg/m³ (200-L sample volume)
Ethylidene dichloride	75-34-3	VOC	TO-14A TO-15 R-3	NIOSH 1003	OSHA 1160	TO-14A: 0.1 ppbv [1003]: 0.01 mg/sample	
Formaldehyde	*50-00-0*	*VVOC*	*TO-11A*	*OSHA 52* *NIOSH 2539* *NIOSH 3500* *NIOSH 5700* *NIOSH 2541* *NIOSH 2016*		*TO-11A: 1 ppbv* *[3500]: 0.5 µg/sample* *[5700]: 0.08 µg/sample* *[2541]: 1 µg/sample* *[2016]: 0.09 µg/sample* *[52]: 16 ppb (20 µg/m³)*	
Heptachlor	76-44-8	SVOC	TO-10A R-29 R-30 R-27		OSHA 1369	TO-10A: 0.01–50 µg/m³ R-29: 0.04–0.1 pg/m³ R-30: 1 ng/m³ [1369]: 0.43 pg/injection	
Hexachlorobenzene	*118-74-1*	*SVOC*	*TO-10A* *R-29* *R-28*		*OSHA 1376*	*TO-10A: 0.01–50 µg/m³* *R-29: 0.04–0.1 pg/m³*	
Hexachlorobutadiene	87-68-3	VOC	TO-14A TO-15 R-3	NIOSH 2543	OSHA H109	TO-14A: 0.1 ppbv [2543]: 0.02 µg/sample	

Compound	CAS	Type	Methods			Detection limit	Notes
1,2,3,4,5,6-Hexachlorocyclohexane (all stereo isomers, including Lindane)	N/A	SVOC	TO-10A R-28 R-29 R-27	NIOSH 5502 R-30		TO-10A: 0.01–50 μg/m³ (g-BHC) R-30: 1 ng/m³ R-29: < 5 pg/m³ [5502]: 3 μg/sample	
Hexachlorocyclopentadiene	77-47-4	SVOC	TO-10A	NIOSH 2518		TO-10A: 0.01–50 μg/m³ [2518]: 5 ng/sample	TO-14A indicated by analogy with VOCs having similar properties
Hexachloroethane	67-72-1	VOC	TO-15	NIOSH 1003 TO-14A TO-3 TO-15	OSHA 1372	[1003]: 0.01 mg/sample	
Hexamethylene diisocyanate	822-06-0	SVOC		OSHA 42 NIOSH 5522 NIOSH 5521 R-23	R-4 [837] R-62	[42]: 2.3 μg/m³ R-23: 1 μg/m³ [5522]: 0.2 μg/sample [5521]: 0.1 μg diisocyanate/sample	[42]: sample volume = 15 L
Hexamethylphosphoramide	680-31-9	SVOC			OSHA H129		
Hexane	110-54-3	VOC	TO-14A TO-15 TO-17 R-6	NIOSH 1500 NIOSH 2549		TO-14A: 0.1 ppbv TO-17: ≤ 0.5 ppb R-6: 0.03 ppbv [2549]:	TO-14A by analogy to other VOCs with similar properties on TO-14A list
Hydrazine	*302-01-2*	*VINC*		*OSHA 108 OSHA 20 NIOSH 3503 R-22, R-84*	*R-55*	*[20]: 1.2 ppbv R-22: 4 ppbv [3503]: 0.9 μg/sample [108]: 0.076 μg/m³*	*[3503]: working range = 0.07–3 ppmv (100-L sample volume) [20]: sample volume = 20 L R-22: sample volume = 2 L*
Hydrochloric acid (Hydrogen chloride)	7647-01-0	VINC	R-19	NIOSH 7903 OSHA ID174SG		R-19: 0.22 ppbv	[7903]: working range = 0.0066–3.3 ppmv (50-L sample volume)

TABLE A3.1

Results of the Survey of Ambient Air Measurement Methods for the 188 HAPs (Chemicals shown in italics are high priority urban HAPS)

Compound	CAS No.	Compound Class[a]	Ambient Measurement Method			Limit of Detection	Comment
			Applicable	Likely	Potential		
Hydrogen fluoride (Hydrofluoric acid)	7664-39-3	VVINC	R-20	R-4[809/205] NIOSH 7903 NIOSH 7902 NIOSH 7906		R-20: 0.08 ppbv [7902]: 3 µg F–/sample [7906]: 3 µg F–/sample (gas); 120 µg F–/sample (particulate)	[7903]: working range = 0.012–6.02 ppmv (50-L sample volume)
Hydroquinone	123-31-9	SVOC		NIOSH 5004	OSHA 1490	[5004]: 0.01 mg/sample	[5004]: working range = 2–25 mg/m³ (30-L sample volume)
Isophorone	78-59-1	VOC	TO-15	NIOSH 2508	OSHA 1538	[2508]: 0.02 mg/sample	[2508]: working range = 0.35–70 ppmv (12-L sample volume)
Maleic anhydride	108-31-6	SVOC	TO-17	OSHA 25 OSHA 86 NIOSH 3512		TO-17: ≤ 0.5 ppb [25]: 0.005 mg/m³ [86]: 33 µg/m³ [3512]: 15 µg/sample	[25]: sample volume = 20 L [86]: sample volume = 60 L
Methanol	67-56-1	VOC	TO-15 TO-17 R-1 R-3	NIOSH 2549 NIOSH 2000 R-64		R-1: 1 ppbv TO-17: ≤ 0.5 ppb [2000]: 0.7 µg/sample	
Methoxychlor	72-43-5	SVOC	TO-10A R-27 R-29		OSHA 1646	TO-10A: 0.01–50 µg/m³ R-27: 1–8 ng/m³ R-29: < 5 pg/m³	

Compound	CAS	Class	Methods	Methods	Methods	Detection limits	Notes
Methyl bromide (Bromomethane)	74-83-9	VVOC	TO-14A TO-15 R-3	NIOSH 2520	OSHA 1680	TO-14A: 0.1 ppbv [2520]: 0.01 mg/sample	
Methyl chloride (Chloromethane)	74-87-3	VVOC	TO-14A TO-15 R-3	NIOSH 1001		TO-14A: 0.1 ppbv [1001]: 0.01 mg/sample	
Methyl chloroform (1,1,1-Trichloroethane)	71-55-6	VOC	TO-14A TO-15 R-3 R-6	OSHA 14 NIOSH 2549		TO-14A: 0.1 ppbv [14]: 0.4 mg/m^3 (0.07 ppm)	
Methyl ethyl ketone (2-Butanone)	78-93-3	VOC	TO-11A TO-15 TO-17 R-1 R-3	OSHA 16 OSHA 84 NIOSH 2549 NIOSH 2500 R-58		R-1: 0.2 ppbv TO-17: ≤ 0.5 ppb TO-11A: 1 ppbv [2500]: 0.004 mg/sample [16]: 1.4 ppm (4.0 mg/m^3)	
Methylhydrazine	60-34-4	VOC		NIOSH S149 R-22 R-84	OSHA 1794 R-55	R-22: 4 ppbv	[S149]: working range = 0.018–0.55 ppmv (20-L sample volume) R-22: sample volume = 2 L
Methyl iodide (Iodomethane)	74-88-4	VVOC	TO-15	NIOSH 1014 TO-14A		[1014]: 0.01 mg/sample TO-14A: 0.1 ppbv	[1014]: working range = 1.7–16.9 ppmv (50-L sample volume)
Methyl isobutyl ketone (Hexone)	108-10-1	VOC	TO-15 TO-17 TO-11A	NIOSH 2549 NIOSH 1300 R-4 [14] R-1 R-58		[14]: 10 ppmv R-58: < 1ppbv TO-17: ≤0.5 ppb [1300]: 0.02 mg/sample TO-11A: 1 ppbv	[1300]: measurement range = 2.1–8.3 mg/sample (1–10-L sample volumes) R-1 suggested by similarity of properties with methyl ethyl ketone
Methyl isocyanate	624-83-9	VOC		OSHA 54	R-62	[54]: 1.9 ppbv (4.8 µg/m^3)	

TABLE A3.1
Results of the Survey of Ambient Air Measurement Methods for the 188 HAPs (Chemicals shown in italics are high priority urban HAPS)

Compound	CAS No.	Compound Class[a]	Ambient Measurement Method			Limit of Detection	Comment
			Applicable	Likely	Potential		
Methyl methacrylate	80-62-6	VOC	TO-15 TO-17	OSHA 94 NIOSH 2537 R-4 [14] R-1 R-3		[14]: 1 ppmv TO-17: ≤ 0.5 ppb [2537]: 0.01 mg/sample [94]: 617 µg/m³	R-1 and R-3 indicated by similarity of properties with ethyl acrylate
Methyl tert-butyl ether	1634-04-4	VOC	TO-15 TO-17 R-1 R-3	NIOSH 1615 R-64		R-1: 1 ppbv TO-17: ≤ 0.5 ppb [1615]: 0.02 mg/sample	
4,4'-Methylenebis-(2-chloroaniline)	101-14-4	NVOC		OSHA 24 OSHA 71		[24]: 3.6 fg/m³ [71]: 440 ng/m³	
Methylene chloride (Dichloromethane)	*75-09-2*	*VOC*	*TO-14A TO-15 TO-17 R-3*	*OSHA 59 OSHA 80 NIOSH 1005 NIOSH 2549*		*TO-14A: 0.1 ppbv TO-17: #0.5 ppb [1005]: 0.4 µg/sample [59]: 29 ppb [80]: 0.697 µg/m³*	
4,4'-Methylenediphenyl diisocyanate	101-68-8	SVOC		OSHA 18 OSHA 47 NIOSH 5522 R-23 R-4 [831]	R-62	[831]: 3–1040 µg/m³ R-23: 1 µg/m³ [5522]: 0.3 µg/sample [18]: 1 µg/m³ (0.10 ppb) [47]: 0.8 µg/m³	[831]: sample volume = 20 L

Compound	CAS	Class	Method	Method	Method	Detection limit	Notes
4,4'-Methylenedianiline	101-77-9	NVOC		OSHA 57 NIOSH 5029		[57]: 81 ng/m³ [5029]: 0.12 -1.2 µg/sample	[57]: sample volume = 100 L [5029]: working range = 0.0002–10 mg/m³ (100-L sample volume)
Naphthalene	91-20-3	SVOC	TO-13A R-7	OSHA 35 NIOSH 1501 NIOSH 5506		TO-13A: <100 pg/m³ [1501]: 0.001 to 0.01 mg/sample with capillary column [35]: 0.4 mg/m3 (0.08 ppm)	R-7: range of measured ambient concentrations = 5.5–182 ng/m³ Volatility presents collection problems with PUF/XAD at high sample volume
Nitrobenzene	98-95-3	VOC	TO-15 TO-17	NIOSH 2005 NIOSH 2017 R-53	OSHA 1870	TO-17: ≤ 0.5 ppb	[2005]: working range = 0.59–2.34 ppmv (55-L sample volume) R-53 by analogy to nitrotoluenes
4-Nitrobiphenyl	92-93-3	SVOC	R-53	R-52	OSHA 1875	R-53: 0.01 ng/m³	R-52 reports ambient data for 3-nitrobipheny l
4-Nitrophenol	100-02-7	SVOC	TO-8 R-53	R-2 R-54	OSHA N607	TO-8: 1-5 ppbv R-53: 0.04 ng/m³ [N607]: 0.003 mg/m³	R-2 by analogy with 2- and 3-nitrophenol R-54 by analogy with 2-nitrophenol
2-Nitropropane	79-46-9	VOC	TO-15	OSHA 15 OSHA 46 NIOSH 2528 R-4 [14]		[14]: 10 ppmv [2528]: 1 µg/sample [46]: 91 µg/m³ [15]: 0.98 mg/m³	[2528]: working range = 1.4–27 ppmv (2-L sample volume)
N-Nitroso-N-methylurea	684-93-5	VOC			TO-7	TO-7: < 0.32 ppbv	Based on similarity of properties with N-nitroso-dimethylamine

TABLE A3.1
Results of the Survey of Ambient Air Measurement Methods for the 188 HAPs (Chemicals shown in italics are high priority urban HAPS)

Compound	CAS No.	Compound Class[a]	Ambient Measurement Method			Limit of Detection	Comment
			Applicable	Likely	Potential		
N-Nitrosodimethylamine	62-75-9	VOC	TO-7	OSHA 27 NIOSH 2522		TO-7: < 0.32 ppbv [2522]: 0.05 µg/sample [27]: 0.13 µg/m³	
N-Nitrosomorpholine	59-89-2	VOC	TO-7	OSHA 17 OSHA 27		TO-7: < 0.32 ppbv [27]: 0.20 µg/m³ [17]: 0.6 µg/m³	Based on similarity of properties with N-Nitroso-dimethylamine
Parathion	56-38-2	SVOC	TO-10A	OSHA 62 NIOSH 5600 R-4 [835]		TO-10A: 0.01-50 µg/m³ [62]: 3.1 µg/m3 (0.26 ppb)	
Pentachloronitrobenzene (Quintobenzene)	82-68-8	SVOC	TO-10A	R-48 R-49		TO-10A: 0.01-50 µg/m³ R-48: 0.1-10 ng/m³ R-49: 10.7-1560 ng/m³	R-48 and R-49: measurement range shown as LOD
Pentachlorophenol	87-86-5	SVOC	TO-10A	OSHA 39 NIOSH 5512 R-50 R-61	R-3	TO-10A: 0.01-50 µg/m R-3: 0.2µg/m³ R-50: < 1 ng/m³ [5512]: 8 µg/sample [39]: 0.007 mg/m³	Use of TO-10A would require filter for particulate material
Phenol	108-95-2	VOC	TO-8 TO-17 R-2 R-54	OSHA 32 NIOSH 2549 NIOSH 2546 R-25 R-60	R-53	TO-8: 1-5 ppbv R-2: 0.02 ppbv R-54: 56–110 pptv TO-17: ≤ 0.5 ppb [2546]: 1 to 3 µg/sample [32]: 0.041 mg/m³ (0.01 ppm)	R-54: LOD shown is range of ambient data.

Compound	CAS	Category	Method	Method	Additional	Detection limit	Notes
p-Phenylenediamine	106-50-3	SVOC		OSHA 87		[87]: 0.44 µg/m³	[87]: sample volume = 100 L)
Phosgene	75-44-5	VVOC	TO-15	OSHA 61, R-81		0.014 mg/m³ (3.5 ppb)	
Phosphine	7803-51-2	VVINC	R-33	OSHA ID180 NIOSH 6002		R-33: 0.15 ppbv; [ID180]: 9 ppbv [6002]: 0.1 fg/sample [180]: 0.015 ppm for 36-L air sample	R-33: LOD derived from reference abstract [ID180]: sample volume = 36 L
Phosphorus	7723-14-0	SVINC		NIOSH 7300 NIOSH 7905 R-77		[7300]: 1 µg/sample [7905]: 0.005 µg/sample	[7300]: working range = 0.005–2 mg/m³ (500-L sample volume) [7905]: working range = 0.04–0.8 mg/m3 (12-L sample volume)
Phthalic anhydride	85-44-9	SVOC		NIOSH S179 OSHA 90	R-26 R-28 R-9	[S179]: 1–36 ng/m³ [90]: 0.048 mg/m³	[S179]: sample volume = 100 L (measurement range shown as LOD) [90]: sample volume = 75 L
Polychlorinated biphenyl (Aroclors)	1336-36-3	*SVOC*	*TO-10A R-29 R-28*	*NIOSH 5503*	*OSHA C107*	*TO-10A: 0.01-50 µg/m³ R-29: 0.04–0.1 pg/m³ [5503]: 0.03 µg/sample*	*Note: Higher chlorinated species, up to decachloro, are probably NVOC*
1,3-Propane sultone	1120-71-4	VOC		R-41			
beta-Propiolactone	57-57-8	VOC	TO-15	R-40		R-40: 1.2 pptv	R-40: not applied to ambient air analysis
Propionaldehyde	123-38-6	VOC	TO-11A	NIOSH 2539		TO-11A: 1 ppbv	

TABLE A3.1
Results of the Survey of Ambient Air Measurement Methods for the 188 HAPs (Chemicals shown in italics are high priority urban HAPS)

Compound	CAS No.	Compound Class[a]	Ambient Measurement Method			Limit of Detection	Comment
			Applicable	Likely	Potential		
Propoxur (Baygon)	114-26-1	SVOC	TO-10A R-27		OSHA 0318	TO-10A: 0.01–50 µg/m³ [0318]: 1 ng/injection	R-27: LOD shown is range of reported ambient data
Propylene dichloride (1,2-Dichloropropane)	*78-87-5*	*VOC*	*TO-14A TO-15 R-3*	*NIOSH 1013*	*OSHA 2190*	*TO-14A: 0.1 ppbv [1013]: 0.1 µg/sample*	
Propylene oxide	75-56-9	VVOC	TO-15	OSHA 88 NIOSH 1612 R-1 R-3 R-13		R-1: 1 ppbv [1612]: 0.01 mg/sample [88]: 83 fg/m³	R-1,R-3, and R-13 indicated by similarity of properties with ethylene oxide
1,2-Propylenimine (2-Methylaziridine)	75-55-8	VOC			OSHA 2213		
Quinoline	*91-22-5*	*SVOC*		*R-14 R-57*			
Quinone (p-Benzoquinone)	106-51-4	SVOC		NIOSH S181	OSHA 2222 R-57	[S181]: 0.17–0.75 mg/m³	[S181]: sample volume = 24 L measurement range shown as LOD)

Compound	CAS	Class	Methods	Methods	Methods	Detection limits	Notes
Styrene	100-42-5	VOC	TO-14A TO-15 TO-17 R-3 R-6	OSHA 9 OSHA 89 NIOSH 1501		TO-14A: 0.1 ppbv TO-17: ≤ 0.5 ppb [1501]: 0.001 to 0.01 mg/sample with capillary column [9]: 0.47 mg/m³ [89]: 426 µg/m³	
Styrene oxide	96-09-3	VOC	TO-15	R-40	OSHA E230 R-3 NIOSH 1614	R-40: 0.41 pptv	R-40: not applied to ambient air analysis R-3, [1614]: Based on comparison of properties with ethylene oxide and 1,2-epoxybutane
2,3,7,8-Tetrachlorodibenzo-p-dioxin	1746-01-6	SVOC	TO-9A R-5 R-51		OSHA 2326	TO-9A: 1-5 pg/m³ R-5: 0.02 pg/m³ R-51: < 0.01 pg/m³	
1,1,2,2-Tetrachloroethane	79-34-5	VOC	TO-14A TO-15 TO-17 R-3	NIOSH 1019	OSHA 2340	TO-14A: 0.1 ppbv TO-17: ≤0.5 ppb [1019]: 0.01 mg/sample	
Tetrachloroethylene (Perchloroethylene)	127-18-4	VOC	TO-14A TO-15 TO-17 R-3 R-6	NIOSH 1003		TO-14A: 0.1 ppbv TO-17: ≤ 0.5 ppb [1003]: 0.01 mg/sample	
Titanium tetrachloride	7550-45-0	VINC					Hydrolyzes extremely rapidly in the ambient atmosphere

TABLE A3.1
Results of the Survey of Ambient Air Measurement Methods for the 188 HAPs (Chemicals shown in italics are high priority urban HAPS)

Compound	CAS No.	Compound Class[a]	Ambient Measurement Method			Limit of Detection	Comment
			Applicable	Likely	Potential		
Toluene	108-88-3	VOC	TO-14A TO-15 TO-17 R-1 R-3 R-6	OSHA 111 NIOSH 1500 NIOSH 1501 NIOSH 2549 NIOSH 4000		TO-14A: 0.1 ppbv R-1: 0.2 ppbv TO-17: ≤ 0.5 ppb [1501]: 0.001 to 0.01 mg/sample with capillary column [4000]: 0.01 mg/sample [111]: 68.3 µg/m³ (charcoal tubes)	
Toluene-2,4-diamine	95-80-7	SVOC		OSHA 65 NIOSH 5516		[5516]: 0.1 µg/sample [65]: 58 fg/m³	[5516]: working range = 3–30 µg/m³ (100-L sample volume)
2,4-Toluene diisocyanate	584-84-9	SVOC		OSHA 18 OSHA 42 NIOSH 5521 NIOSH 2535 NIOSH 5522 R-4 [837] R-23 R-24	R-62	R-23: 1 fg/m³ R-24: 7.24 fg/m³ [837]: 0.05–1.01 mg/m³ [5522]: 0.1 µg/sample [5521]: 0.1 µg diisocyanate/sample [2535]: 0.1 µg/sample [18]: 1 fg/m3 (0.15 ppb)	[2535]: working range = 0.03–2.5 mg/m3 (10-L sample volume) [837]: sample volume = 20 L
o-Toluidine	95-53-4	SVOC		OSHA 73 NIOSH 2017 NIOSH 2002		[73]: 0.97 µg/m³	[2002]: working range = 5–60 mg/m3 (55-L sample volume) [73]: sample volume = 100 L

Compound	CAS	Class	Methods			Detection Limits	Comments
Toxaphene (chlorinated camphene)	8001-35-2	SVOC	R-31 R-28	NIOSH S67	OSHA 0612	R-31: < 0.1 ng/m3	[S67]: working range = 0.05–1.5 mg/m3 (15-L sample volume) R-31: LOD estimated
1,2,4-Trichlorobenzene	120-82-1	VOC	TO-14A TO-15 R-3	NIOSH 5517		TO-14A: 0.1 ppbv [5517]: 0.001 µg/mL in hexane	
1,1,2-Trichloroethane	79-00-5	VOC	TO-14A TO-15 TO-17 R-3	OSHA 11 NIOSH 1003		TO-14A: 0.1 ppbv TO-17: ≤ 0.5 ppb [1003]: 0.01 mg/sample [11]: 0.14 mg/m^3	
Trichloroethylene	*79-01-6*	*VOC*	*TO-14A TO-15 R-6*	*NIOSH 3701*		*TO-14A: 0.1 ppbv [3701]: 0.25 ng per 1-mL injection*	
2,4,5-Trichlorophenol	95-95-4	SVOC	TO-10A R-50	R-54	R-3	TO-10A: 0.01–50 µg/m^3 R-50: 0.07 ng/m^3	Use of TO-10A would require filter for particulate-phase material R-54 reports sum of 2,4,5- and 2,4,6- isomers
2,4,6-Trichlorophenol	88-06-2	SVOC	TO-10A R-50	R-54	R-3	TO-10A: 0.01–50 µg/m^3 R-3: 0.2 µg/m^3 R-50: 0.07 ng/m^3	TO-10A by analogy with 2,4,5-TCP. Use of TO-10A would require filter for particulate material R-54 reports sum of 2,4,5- and 2,4,6- isomers

TABLE A3.1
Results of the Survey of Ambient Air Measurement Methods for the 188 HAPs (Chemicals shown in italics are high priority urban HAPS)

Compound	CAS No.	Compound Class[a]	Ambient Measurement Method			Limit of Detection	Comment
			Applicable	Likely	Potential		
Triethylamine	121-44-8	VOC		NIOSH S152 R-9	OSHA 2480	[S152]: 2–71 ppbv	[S152]: sample volume = 100 L (measurement range shown as LOD) R-9 indicated by comparison of properties with dimethylformamide
Trifluralin	1582-09-8	SVOC	TO-10A R-29		OSHA T338	TO-10A: 0.01-50 µg/m³ R-29: < 100 pg/m³	
2,2,4-Trimethylpentane	540-84-1	VOC	TO-14A TO-15 R-6			TO-14A: 0.1 ppbv R-6: 0.025 ppbv	TO-14A and TO-15 indicated by analogy with other VOCs having similar properties, and based on canister stability data.
Vinyl acetate	108-05-4	VOC	TO-15 R-1 R-3	OSHA 51 NIOSH 1453		R-1: 1 ppbv [1453]: 1 µg/sample [51]: 0.04 mg/m³	
Vinyl bromide	593-60-2	VVOC	TO-15	OSHA 8 NIOSH 1009 TO-14A		[1009]: 3 µg/sample [8]: 0.2 ppm	TO-14A indicated by analogy to other VVOCs with similar properties

Name	CAS	Category	Methods	Methods	Detection Limits
Vinyl chloride	75-01-4	VVOC	TO-14A, TO-15, R-3	OSHA 4, OSHA 75, NIOSH 1007	TO-14A: 0.1 ppbv; [1007]: 0.04 µg/sample; [4]: 0.25 ppm; [75]: 0.051 mg/m³
Vinylidene chloride (1,1-Dichloroethylene)	75-35-4	VVOC	TO-14A, TO-15, R-3	OSHA 19, NIOSH 1015	TO-14A: 0.1 ppbv; [1015]: 7 µg/sample; [19]: 0.2 mg/m³
Xylenes (mixed isomers)	1330-20-7	VOC	TO-14A, TO-15, TO-17, R-3	NIOSH 1501, NIOSH 2549	TO-14A: 0.1 ppbv; TO-17: ≤ 0.5 ppb; [1501]: 0.001 to 0.01 mg/sample with capillary column
o-Xylene	95-47-6	VOC	TO-14A, TO-15, TO-17, R-3, R-6	NIOSH 1501, NIOSH 2549	TO-14A: 0.1 ppbv; TO-17:≤ 0.5 ppb; [1501]: 0.001 to 0.01 mg/sample with capillary column
m-Xylene	108-38-3	VOC	TO-14A, TO-15, TO-17, R-6	NIOSH 1501, NIOSH 2549	TO-14A: 0.1 ppbv; TO-17:≤ 0.5 ppb; [1501]: 0.001 to 0.01 mg/sample with capillary column
p-Xylene	106-42-3	VOC	TO-14A, TO-15, TO-17, R-6	NIOSH 1501, NIOSH 2549	TO-14A: 0.1 ppbv; TO-17: ≤ 0.5 ppb; [1501]: 0.001 to 0.01 mg/sample with capillary column
Antimony Compounds		NVINC	IO-3, R-4 [301]	OSHA ID125g, R-4 [804 & 822B]	[301]: 0.05 µg/m³; [804]: 4 µg/m³; [822B]²: 17.5 ng/m³

[301]: sample volume = 20 m³

[804]: sample volume = 60 L

TABLE A3.1
Results of the Survey of Ambient Air Measurement Methods for the 188 HAPs (Chemicals shown in italics are high priority urban HAPs)

Compound	CAS No.	Compound Class[a]	Ambient Measurement Method			Limit of Detection	Comment
			Applicable	Likely	Potential		
Arsenic Compounds (Inorganic including arsine)		VVINC/ NVINC	IO-3 R-4 [302]	OSHA ID125g NIOSH 6001 NIOSH 7300 NIOSH 7900 NIOSH 7901 NIOSH 79300 R-34 R-4 [804 & 822B]		R-34: 1 ppbv [302]: 0.1 µg/m³ [804]: 0.4 µg/m³ [822B]²: 10 ng/m³ [7900]: 0.02 µg/sample [7901]: 0.06 µg/sample [6001]: 0.004 µg/sample	[6001]: working range = 0.3–62 ppbv (10-L sample volume) R-34: chemiluminescence methods [302]: sample volume = 10 m³ [804]: sample volume = 60 L
Beryllium Compounds		NVINC	IO-3 R-4 [822]	OSHA ID125g NIOSH 7102 NIOSH 7300		[822]: 0.08 µg/m³ [7102]: 0.004 µg/sample	[822]: sample volume = 0.24 m³
Cadmium Compounds		NVINC	IO-3 R-4 [822]	OSHA ID125g NIOSH 7300 NIOSH 7048 R-4 [822B]		[822]: 0.03 µg/m³ [822B]²: 22.5 ng/m³ [7048]: 0.05 µg/sample	[822]: sample volume = 0.24 m³
Chromium Compounds		NVINC	IO-3 R-4 [822]	OSHA ID125g NIOSH 7024 NIOSH 7300 R-4 [822B]		[822]: 0.2 µg/m³ [822B]²: 25 ng/m³ [7024]: 0.06 µg/sample	[822]: sample volume = 0.24 m³
Cobalt Compounds		NVINC	IO-3 R-4 [822]	OSHA ID125g NIOSH 7300 NIOSH 7027 R-4 [822B]		[822]: 0.3 µg/m³ [822B]²: 12.5 ng/m³ [7027]: 0.6 µg/sample	[822]: sample volume = 0.24 m³

Compound	Type	Method	Method	Values / LOD	Notes
Coke Oven Emissions: represented by: Naphthalene and Coronene (see also VOCs, e.g., benzene, toluene, xylene)	SVOC/ NVOC	TO-13A R-7 R-14	OSHA 35 OSHA 58 NIOSH 1501 NIOSH 5506 R-4 [836]	TO-13A: <100 pg/m³ R-14: 0.09–1.4 ng/m³ [1501]: 0.001 to 0.01 mg/sample with capillary column	R-7: range of measured ambient concentrations = 5.5–182 ng/m³ R-14: LOD shown is range of measured ambient concentrations
Cyanide Compounds: Hydrogen Cyanide and Particulate Cyanide	VVINC/ NVINC	NIOSH 6010 R-35 NIOSH 7904		[7904]: 2.5 µg CN– [6010]: 1 µg CN–	[6010]: working range = 0.893–232 ppmv (3-L sample volume) R-35: 1 pg quantitation limit [7904]: working range (as CN–) = 0.5–15 mg/m³ (sample volume = 90 L)
Glycol ethers	SVOC	TO-17	R-45	R-45: > 0.74 mg/m³ TO-17: ≤ 0.5 ppb	See also ethylene glycol R-45: developed for 9 different glycol ethers, LOD calculated for ethylene glycol monobutyl ether
Lead Compounds	NVINC	IO-3 R-4 [822]	OSHA ID125g NIOSH 7700 NIOSH 7300 NIOSH 7105 NIOSH 7082 NIOSH 7702 NIOSH 7701 R-4 [822B]	[822]: 0.8 fg/1m³ [822B]: 10 ng/m³ [7105]: 0.02 µg/sample [7082]: 2.6 µg/sample [7702]: 6 µg/sample Pb [7701]: 0.09 µg/sample	[822]: sample volume = 0.24 m³
Manganese Compounds	NVINC	IO-3 R-4 [822]	OSHA ID125g NIOSH 7300 R-4 [822B]	[822]: 0.1 µg/m³ [822B]: 20 ng/m³	[822]: sample volume = 0.24 m³

TABLE A3.1
Results of the Survey of Ambient Air Measurement Methods for the 188 HAPs (Chemicals shown in italics are high priority urban HAPS)

Compound	CAS No.	Compound Class[a]	Ambient Measurement Method			Limit of Detection	Comment
			Applicable	Likely	Potential		
Mercury Compounds		*SVINC/ NVINC*	*IO-3 IO-5* *R-4 [317]* *R-18*	*OSHA ID140* *OSHA ID125g* *OSHA ID145* *NIOSH 6009* *R-4 [822B]*		*[317]: 0.002–0.06 ppbv* *R-18: 0.001–0.06 pptv* *[6009]: 0.03 µg/sample* *[ID145]: 0.002 mg/m³ for 10-L air sample*	
Fine Mineral Fibers		NVINC		NIOSH 7400 R-21		[7400]: 7 fibers/mm² filter area	[7400]: working range= 0.04–0.5 fiber/cc (1000-L sample volume)
Nickel Compounds		*NVINC*	*IO-3* *R-4 [822]*	*OSHA ID125g* *NIOSH 7300* *R-4 [822B]*		*[822]: 0.3 µg/m³* *[822B]: 10 ng/m³*	*[822]: sample volume = 0.24 m³*
Polycyclic Organic Matter: represented by Naphthalene and Coronene		*SVOC/ NVOC*	*TO-13A* *R-7* *R-14*	*OSHA 35* *NIOSH 1501* *NIOSH 5506* *R-4 [836]*		*TO-13A: <100 pg/m³* *R-14: 0.09–1.4 ng/m³* *[1501]: 0.001 to 0.01 mg/sample with capillary column*	*R-7: range of measured ambient concentrations = 5.5–182 ng/m³* *R-14: LOD shown is range of measured ambient concentrations*
Radionuclides (including radon)		VVINC/ NVINC/ VINC	R-4 [606, A/B] R-16 R-17	OSHA 2560	OSHA R100	[606B]: 0.1 pCi/L Varies depending on radionuclide	Long-lived radionuclides in air include: ^{222}Rn, ^{220}Rn, ^{131}I, ^{85}Kr, ^{3}H (gases); ^{210}Po, ^{210}Pb, ^{7}Be, ^{239}Pn, ^{238}Pn, ^{144}Ce, ^{137}Cs, ^{90}Sr (particles)

Selenium Compounds	NVINC	IO-3	OSHA ID125g NIOSH 7300 R-4 [804 & 822B]	[804]: 0.4 µg/m³; [822B]²: 10 ng/m³	[804]: sample volume= 60 L

a Compound class description is defined in Table 2.1.

Methods Cited in Table 3.1

EPA TO-Compendium Methods

TO-1   Method for the Determination of Volatile Organic Compounds in Ambient Air Using Tenax Adsorption and Gas Chromatography/Mass Spectrometry (GC/MS), Revision 1.0   4/84

TO-2   Method for the Determination of Volatile Organic Compounds in Ambient Air by Carbon Molecular Sieve Adsorption and Gas Chromatography/Mass Spectrometry (GC/MS) Revision 1.0   4/84

TO-3   Method for the Determination of Volatile Organic Compounds in Ambient Air Using Cryogenic Preconcentration Techniques and Gas Chromatography with Flame Ionization and Electron Capture Detection, Revision 1.0   4/84

TO-4A   Determination of Pesticides and Polychlorinated Biphenyls in Ambient Air Using High Volume Polyurethane Foam (PUF) Sampling Followed by Gas Chromatographic/Multi-Detector Detection (GC/MD), Second Edition   1/97

TO-5   Method for the Determination of Aldehydes and Ketones in Ambient Air Using High Performance Liquid Chromatography (HPLC), Revision 1.0   4/84

TO-6   Method for the Determination of Phosgene in Ambient Air Using High Performance Liquid Chromatography (HPLC), Revision 1.0   9/86

TO-7   Method for the Determination of N-Nitrosodimethylamine in Ambient Air Using Gas Chromatography, Revision 1.0   9/86

TO-8   Method for the Determination of Phenol and Methylphenols (Cresols) in Ambient Air Using High Performance Liquid Chromatography, Revision 1.0   9/86

TO-9A   Determination of Polychlorinated, Polybrominated and Brominated/Chlorinated Dibenzo-p-Dioxins (TCDDs) and Dibenzofurans in Ambient Air, Second Edition   1/97

TO-10A   Determination of Pesticides and Polychlorinated Biphenyls In Ambient Air Using Low Volume Polyurethane Foam (PUF) Sampling Followed by Gas Chromatographic/Multi-Detector Detection (GC/MD), Second Edition   1/97

TABLE A3.1 (CONTINUED)
Results of the Survey of Ambient Air Measurement Methods for the 188 HAPs

TO-11A	Determination of Formaldehyde in Ambient Air Using Adsorbent Cartridge Followed by High Performance Liquid Chromatography (HPLC), Active Sampling Methodology, Second Edition	1/97
TO-12	Method for the Determination of Non-Methane Organic Compounds (NMOC) in Ambient Air Using Cryogenic Preconcentration and Direct Flame Ionization Detection (PDFID)	
TO-13A	Determination of Polycyclic Aromatic Hydrocarbons (PAHs) in Ambient Air Using Gas Chromatography/Mass Spectrometry (GC/MS), Second Edition	1/97
TO-14A	Determination of Volatile Organic Compounds (VOCs) in Ambient Air Using Specially Prepared Canisters With Subsequent Analysis By Gas Chromatography, Second Edition	1/97
TO-15	Determination of Volatile Organic Compounds (VOCs) in Air Collected in Specially Prepared Canisters And Analyzed by Gas Chromatography/Mass Spectrometry (GC/MS), Second Edition	1/97
TO-16	Long-Path Open-Path Fourier Transform Infrared Monitoring Of Atmospheric Gases, Second Edition	1/97
TO-17	Determination of Volatile Organic Compounds in Ambient Air Using Active Sampling Onto Sorbent Tubes, Second Edition	1/97
OSHA Methods		
OSHA ID180	Ion Chromatography using glass tube containing potassium hydroxide-coated carbon for Phosphine in Workplace Atmospheres	6/91
OSHA ID160	Fiber counting by Phase Contract Microscopy (PCM) using mixed-cellulose ester filter for Asbestos in Air	7/97
OSHA ID145	Cold Vapor-Atomic Absorption Spectrophotometry using mercury-containing ester filter for Particulate Mercury in Workplace Atmospheres	12/89
OSHA ID140	Cold Vapor-Atomic Absorption Spectrophotometry using solid sorbent sampling device (Hydrar or hopcalite) for Mercury Vapor in Workplace Atmospheres	6/91
OSHA ID125G	Inductively Coupled Plasma- Atomic Emission Spectroscopy using mixed-cellulose ester membrane filter in styrene cassette for Metal and Metalloid Particulates in Workplace Atmospheres	4/91

Method	Description	Date
OSHA ID101	Ion Specific Electrode using a midget fritted glass bubbler containing sulfamic acid for Chlorine in Workplace Atmospheres	5/91
OSHA 95	HPLC- UV Detector using cassette containing two glass fiber filters for Ethylene Thiourea	8/92
OSHA 94	Gas Chromatography- Flame Ionization Detector using glass sampling tubes containing 4-tert-butylcatechol-coated charcoal for Methyl Methacrylate	6/92
OSHA 93	Gas Chromatography- Electron Capture Detector using sampling device consisting of cassettes each containing two sulfuric acid-treated glass fiber filters for 4-Aminobiphenyl	1/92
OSHA 92	Gas Chromatography- Flame Ionization Detector using glass sampling tubes containing 4-tert-butylcatechol-coated charcoal for Ethyl Acrylate and Methyl Acrylate	12/91
OSHA 90	HPLC- UV Detector using sampling device containing two coated glass fiber filters for Phthalic Anhydride	10/91
OSHA 9	Gas Chromatography- Flame Ionization Detector using sorbent tubes (charcoal) for Styrene; Superceded by Method No. 89	2/80
OSHA 89	Gas Chromatography- Flame Ionization Detector using glass sampling tubes containing coated charcoal for Divinylbenzene, Ethylvinylbenzene, and Styrene	7/91
OSHA 88	Gas Chromatography- Flame Ionization Detector using Anasorb 747 sorbent tubes for Propylene Oxide	6/91
OSHA 87	HPLC- UV Detector using cassettes each containing sulfuric acid-treated glass fiber filters for m-, o-, and p-Phenylenediamine	2/91
OSHA 86	HPLC- UV Detector using coated glass fiber filters for Maleic Anhydride	12/90
OSHA 84	Gas Chromatography- Flame Ionization Detector using glass sampling tubes containing Carbosieve S-III adsorbent for 2-Butanone	7/90
OSHA 80	Gas Chromatography- Flame Ionization Detector using glass sampling tubes containing Carbosieve S-III adsorbent for Methylene Chloride	2/90
OSHA 8	Gas Chromatography- Flame Ionization Detector using sorbent tube (charcoal) for Vinyl Bromide	5/79
OSHA 75	Gas Chromatography- Flame Ionization Detector using glass sampling tubes containing Carbosieve S-III adsorbent for Vinyl Chloride	4/89
OSHA 73	Gas Chromatography- Electron Capture Detector using cassettes each containing two sulfuric acid-treated glass fiber filters for o-, m-, p-Toluidene	8/88

TABLE A3.1 (CONTINUED)
Results of the Survey of Ambient Air Measurement Methods for the 188 HAPs

OSHA 71	Gas Chromatography Electron Capture Detector using cassettes each containing two sulfuric acid-treated glass fiber filters for 4,4'-Methylenebis(o-Chloroaniline)	7/89
OSHA 68	Gas Chromatography-Nitrogen Selective Detector using sorbent tubes (XAD-2) for Acetaldehyde	1/88
OSHA 67	Gas Chromatography- Electron Capture Detector using OSHA versatile sampler (OVS-2) tubes containing glass fiber filter and XAD-2 for Chlordane	12/87
OSHA 66	Gas Chromatography- Nitrogen/Phosphorous Detector using sorbent tubes (charcoal) for N,N-Dimethylformamide	8/87
OSHA 65	Gas Chromatography- Electron Capture Detector using sampling device consisting of cassettes each containing two sulfuric acid-treated glass fiber filters for Benzidine, 3,3'-Dichlorobenzidine, 2,4-Toluenediamine, and 2,6-Toluenediamine	7/89
OSHA 63	HPLC- UV Detector using OSHA versatile sampler (OVS-2) tubes containing glass fiber filter and XAD-2 for Carbaryl	3/87
OSHA 62	Gas Chromatography- Flame Photometric Detector (FPD) using glass sampling tubes each containing glass fiber filter and two sections of XAD-2 adsorbent for Chlorpyrifos, DDVP (Dichlorvos), Diazinon, Malathion, and Parathion	10/86
OSHA 61	Gas Chromatography- Nitrogen Selective Detector using coated XAD-2 sorbent tubes for Phosgene	8/86
OSHA 59	Gas Chromatography- Flame Ionization Detector using special sorbent tubes (charcoal) for Methylene Chloride; Superceded by Method No. 80 when a standard size adsorbent tube is desired	4/86
OSHA 57	Gas Chromatography- Electron Capture Detector using sulfuric acid-treated glass fiber filters for 4,4'-Methylenedianiline (MDA)	7/89
OSHA 56	Gas Chromatography- Flame Ionization Detector using sorbent tubes (charcoal) for 1,3-Butadiene	12/85
OSHA 54	HPLC- UV or Fluorescence Detector using XAD-7 tubes coated with 1-2PP for Methyl Isocyanate (MIC)	4/85
OSHA 52	Gas Chromatography- Nitrogen Selective Detector using sorbent tubes (XAD-2) for Acrolein and Formaldehyde	6/89
OSHA 51	Gas Chromatography- Flame Ionization Detector using Ambersorb XE-347 sampling tubes for Vinyl Acetate	3/85

Method	Description	Date
OSHA 50	Gas Chromatography- Electron Capture Detector using hydrobromic acid-coated charcoal tubes for Ethylene Oxide	1/85
OSHA 5	Gas Chromatography- Flame Ionization Detector using sorbent tube (charcoal) for Chloroform	5/79
OSHA 49	Gas Chromatography- Electron Capture Detector using 3M Ethylene Oxide Monitors (passive monitor) for Ethylene Oxide	11/84
OSHA 47	HPLC- UV or Fluorescence Detector using (1-2PP)-coated glass fiber filter for Methylene Bisphenyl Isocyanate (MDI)	3/89
OSHA 46	Gas Chromatography- Flame Ionization Detector using XAD-4 sampling tubes for 1-Nitropropane and 2-Nitropropane	1/84
OSHA 44	Gas Chromatography using a Thermal Energy Analyzer (TEA) equipped with an explosives analysis package (EAP) using a modified commercial Tenax resin tubes for 2,4-Dinitrotoluene	10/83
OSHA 42	HPLC- UV or Fluorescence Detector using glass fiber filters coated with 1-2PP in open-face cassettes for Diisocyanates: 1,6-Hexamethylrene Diisocyanate (HDI), Toluene-2,6-Diisocyanate (2,6-TDI), and Toluene-2,4-Diisocyanate (2,4-TDI)	3/89
OSHA 4	Gas Chromatography- Flame Ionization Detector using sorbent tube (charcoal) for Vinyl Chloride; Superceded by Method No. 75	4/79
OSHA 39	HPLC- UV Detector using XAD-7 sorbent tubes in series for Pentachlorophenol	10/82
OSHA 37	Gas Chromatography- Nitrogen/Phosphorous Detector using sorbent tubes (charcoal) for Acrylonitrile	5/82
OSHA 35	Gas Chromatography- Flame Ionization Detector using Chromosorb 106 tubes for Naphthalene	4/82
OSHA 32	HPLC- UV Detector using XAD-7 sampling tube for Phenol and Cresol	11/81
OSHA 30	Gas Chromatography- Electron Capture Detector using two charcoal tubes in a series for Ethylene Oxide; Superceded by Method No. 50	8/81
OSHA 3	Gas Chromatography- Electron Capture Detector using sorbent tube (charcoal) for Ethylene Dichloride	4/79
OSHA 28	HPLC- UV Detector using two XAD-2 sampling tubes in a series for Acrylic Acid	4/81
OSHA 27	Gas Chromatography- Thermal Energy Analyzer Detector using ThermoSorb/N air samplers for Volatile Nitrosamine: N-Nitrosodimethylamine (NDMA), N-Nitrosodiethylamine (NDEA), N-Nitrosodi-n-Propylamine (NDPA), N-Nitrosodi-n-butylamine (NDBA), N-Nitrosopiperidine (NPIP), N-Nitrosopyrrolidine (NPYR), and N-Notrosomorpholine (NMOR)	2/81

TABLE A3.1 (CONTINUED)
Results of the Survey of Ambient Air Measurement Methods for the 188 HAPs

OSHA 25	Reverse phase HPLC- UV Detector using sorbent tubes (XAD-2) in a series for Maleic Anhydride; Superceded by Method No. 86	2/81
OSHA 24	HPLC- UV Detector using HCl bubbler for 4,4'-Methylenebis(o-Chloroaniline); Superceded by Method No. 71	2/81
OSHA 21	Gas Chromatography- Nitrogen/Phosphorous Detector using glass fiber filter followed by silica gel tube for Acrylamide	10/80
OSHA 20	HPLC using sulfuric acid coated Gas Chrom R for Hydrazine	9/80
OSHA 2	Gas Chromatography- Electron Capture Detector using sorbent tube (charcoal) for Ethylene Dibromide	11/81
OSHA 19	Gas Chromatography- Flame Ionization Detector using sorbent tube (charcoal) for Vinylidene Chloride (1,1-Dichloroethene)	4/80
OSHA 18	HPLC using collection in a bubbler containing nitro reagent in toluene for Diisocyanates; Superceded by Method No. 47	2/80
OSHA 17	Gas Chromatography- Chemiluminescence Detector using two sampling tubes in series for N-Nitrosomorpholine; Superceded by Method No. 27	1/80
OSHA 16	Gas Chromatography- Flame Ionization Detector using collection tubes (silica gel) for 2-Butanone (Methyl Ethyl Ketone); Superceded by Method No. 84	1/80
OSHA 15	Gas Chromatography- Flame Ionization Detector using Chromosorb 106 tubes for 2-Nitropropane; Superceded by Method No. 46	1/80
OSHA 14	Gas Chromatography- Flame Ionization Detector using sorbent tube (charcoal) for 1,1,1-Trichloroethane	1/80
OSHA 12	Gas Chromatography using sorbent tubes (charcoal) desorbed with carbon disulfide for Benzene	8/80
OSHA 112	Gas Chromatography- ECD using sampling tubes (Chromasorb) for β-Chloroprene	
OSHA 111	Gas Chromatography- Flame Ionization Detector using sorbent tubes (charcoal or Anasorb 747) for Toluene	4/98
OSHA 11	Gas Chromatography- Flame Ionization Detector using sorbent tube (charcoal) for 1,1,2-Trichloroethane	2/80
OSHA 108	LC- UV Detector using cassettes each containing two sulfuric acid-treated glass fiber filters for Hydrazine	2/97

Method	Description	Date
OSHA 104	Gas Chromatography- Flame Ionization Detector using OVS-Tenax sampling tubes for Dimethylphthalate, Diethylphthalate, Dibutylphthalate, Di-2-Ethyl Hexylphthalate, and Di-n-Octylphthalate	8/94
OSHA 10	Gas Chromatography- Electron Capture Detector using derivatizing reagent bubblers connected in a series for Bis-Chloromethyl Ether and Chloromethyl Methyl Ether	8/79
OSHA T338	Partially Validated Method- HPLC/UL56 using OSHA Versatile sampler (OVS-2) -13mm XAD-2 tube with glass fiber filter for Trifluralin	
OSHA R220	Gas Chromatography/FPD using graphatized carbon black for Carbonyl Sulfide- Not Validated	
OSHA R100	E-PERM Sampler; passive monitor for Radon	
OSHA N607	HPLC/UL53 using midget impinger containing NaOH for p-Nitrophenol- Not Validated	
OSHA H109	Partially Validated Method- Gas Chromatography ECD using XAD-2 tube for Hexachlorobutadiene (NIOSH 307)	
OSHA E230	Partially Validated Method- Gas Chromatography FID using Tenax GC tube for Styrene Oxide (NIOSH 303)	
OSHA D639	HPLC/U using bulk sample for Dibenzofuran- Not Validated	
OSHA D177	Gas Chromatography FID using charcoal tube for 1,3-Dichloropropene- Not Validated (OSHA Modified NIOSH 1003)	
OSHA D129	Partially Validated Method- HPLC/UL59 using coated XAD-2 tube for Diethanolamine	
OSHA C107	Partially Validated Method- Gas Chromatography ECD using OSHA Versatile sampler (OVS-2) -13mm XAD-2 tube with glass fiber filter for Polychlorinated Biphenyls (Aroclors)	
OSHA B408	Gas Chromatography FID using Tenax GC tube for Benzotrichloride- Not Validated	
OSHA A625	Partially Validated Method- Gas Chromatography with NPD using silica gel tube for Acetamide	
OSHA A169	Partially Validated Method- Gas Chromatography FID using Tenax GC tube for Acetophenone	
OSHA 2480	Partially Validated Method- Gas Chromatography FID using coated XAD-2 tube for Trimethylamine	

TABLE A3.1 (CONTINUED)
Results of the Survey of Ambient Air Measurement Methods for the 188 HAPs

OSHA 2450	Gas Chromatography ECD using sulfuric acid-coated glass fiber filters in 3 piece cassette for 3,3'-Dimethylbenzidine (Fully Validated OSHA 71)
OSHA 2340	Gas Chromatography- FID using petroleum base charcoal tube for 1,1,2,2-Tetrachloroethane
OSHA 2326	Midget impinger containing Xylene for 2,3,7,8- Tetrachlorodibenzo-p-Dioxin- Not Validated
OSHA 2222	HPLC/UL68 using XAD-2 tube for Quinone (Fully Validated Modified NIOSH S181)
OSHA 2213	HPLC/UL53 using midget fritted glass bubbler containing Folins reagent for Propyleneimine- Not Validated
OSHA 2190	Gas Chromatography- FID using charcoal tubes for 1,2-Dichloropropane
OSHA 1911	Partially Validated Method- Gas Chromatography FID using OSHA Versatile sampler (OVS-7) -13mm XAD-7 tube with glass fiber filter for Ethylene Glycol (NIOSH 5523)
OSHA 1875	Partially Validated Method- HPLC/UL56 using OSHA Versatile sampler (OVS-2) -13mm XAD-2 tube with glass fiber filter for 4-Nitrobiphenyl OR Gas Chromatography FID using glass fiber filter in Swinnex J cassette in series with silica gel tube for 4-Nitrobiphenyl – Not Validated (OSHA 273)
OSHA 1870	Gas Chromatography- FID using silica gel tube for Nitrobenzene
OSHA 1794	HPLC/UL52 using sulfuric acid-coated Gas Chrom R tube for Methyl hydrazine- Not Validated
OSHA 1680	Partially Validated Method- Gas Chromatography FID using two Anasorb 747 tubes in series for Methyl Bromide
OSHA 1646	Partially Validated Method- Gas Chromatography ECD using OSHA Versatile sampler (OVS-2) -13mm XAD- tube with glass fiber filter for Methoxychlor
OSHA 1538	Gas Chromatography- FID using petroleum base charcoal tube for Isophorone
OSHA 1490	Gas Chromatography- FID using phosphoric acid- coated XAD-7 for Hydroquinone
OSHA 1376	Gas Chromatography ECD using glass fiber filter for Hexachlorobenzene- Not Validated
OSHA 1372	Gas Chromatography- FID using charcoal tube for Hexachloroethane
OSHA 1369	Partially Validated Method- Gas Chromatography ECD using OSHA Versatile sampler (OVS-2) -13mm XAD-2 tube with glass fiber filter for Heptachlor

OSHA 1160	Gas Chromatography FID using charcoal tube for Ethylidene dichloride (1,1-Dichloroethane) (Fully Validated NIOSH 1003)
OSHA 1110	Gas Chromatography- FID using two charcoal tubes in series for Ethyl Chloride
OSHA 1015	Gas Chromatography FID using OSHA Versatile sampler (OVS Tenax) with glass fiber filter for Bis (2-ethylhexyl)phthalate (Fully Validated OSHA 104)
OSHA 1011	Partially Validated Method- Gas Chromatography FID using XAD-7 tube for Biphenyl
OSHA 1010	Gas Chromatography- FID using charcoal tube for Dioxane
OSHA 0990	Gas Chromatography TEA/EAP using glass fiber filter contained within Tenax-GC tube for 2,4-Dinitrotoluene (Fully Validated OSHA 44)
OSHA 0975	HPLC/UL61 using mixed cellulose ester filter in series with midget impinger for 4,6-Dinitro-o-cresol (Fully Validated NIOSH S-166)
OSHA 0960	Partially Validated Method- Gas Chromatography FPD using Porapak Q tube for Dimethyl Sulfate
OSHA 0940	Colorimetric analysis using midget fritted glass bubbler for 1,1-Dimethylhydrazine (Fully Validated NIOSH S-143)
OSHA 0935	Partially Validated Method- Gas Chromatography ECD using petroleum base charcoal tube for 1,2-Dibromo-3-Chloropropane
OSHA 0931	Partially Validated Method- Gas Chromatography FID using coated XAD-7 tube for N,N-Dimethylaniline
OSHA 0929	HPLC/UL53 using glass fiber filter in series with Midget Impinger for 4-Dimethylaminoazobenzene -Not Validated
OSHA 0913	Gas Chromatography FID using silica gel tube for Diethyl Sulfate -Not Validated
OSHA 0873	Gas Chromatography ECD using coated glass fiber filters for 3,3'-Dimethoxybenzidine (Fully Validated OSHA 71)
OSHA 0861	Gas Chromatography- FID usind XAD-2 coated tube for Diazomethane
OSHA 0850	Gas Chromatography FPD using OSHA Versatile sampler (OVS-2) -13mm XAD-2 tube with glass fiber filter for Dichlorvos (Fully Validated OSHA 62)
OSHA 0760	HPLC/UL33 or Gas Chromatography FID using XAD-7 tube for Cresol (All Isomers) (Fully Validated OSHA 32)
OSHA 0645	Gas Chromatography- FID using charcoal tube for Epichlorohydrin

TABLE A3.1 (CONTINUED)
Results of the Survey of Ambient Air Measurement Methods for the 188 HAPs

OSHA 0618	Partially Validated Method- HPLC/UL66 using two Tenax- GC tubes in series for alpha-Chloroacetophenone (OSHA Modified NIOSH 291)	
OSHA 0612	Gas Chromatography ECD using mixed cellulose ester filter for Toxaphene (Fully Validated NIOSH S-67)	
OSHA 0571	Partially Validated Method- Gas Chromatography FID using XAD-2 for Catechol	
OSHA 0529	Partially Validated Method- HPLC/UL59 using OSHA Versatile sampler (OVS-2) -13mm XAD-2 tube with glass fiber filter for Captan	
OSHA 0400	Gas Chromatography- FID using charcoal tube for Bromoform	
OSHA 0318	Partially Validated Method- HPLC/UL58 using OSHA Versatile sampler (OVS-2) -13mm XAD-2 tube with glass fiber filter for Propoxur	
OSHA 0225	HPLC/UL58 using XAD-2 tube for Anisidine (o,p-Isomers)	
OSHA 0220	Partially Validated Method- Gas Chromatography FID using coated XAD-7 tube for Aniline	
OSHA 0117	Partially Validated Method- HPLC/UL62 using two Anasorb 747 tubes in series for Acrylic Acid	
OSHA 0115	Partially Validated Method- HPLC/UL58 using OSHA Versatile Sampler (OVS-7) for Acrylamide	
OSHA 0065	HPLC/UL53 using glass fiber filter for 2-Acetylaminofluorene- Not Validated	
NIOSH Methods		
NIOSH 9002	Microscopy- Stereo and Polarized Light with Dispersion Staining using bulk sample for Asbestos (Bulk) by PLM	8/94
NIOSH 9000	X-Ray Powder Diffraction using bulk sample for Asbestos, Chrysotile by XRD	8/94
NIOSH 7906	Ion Chromatography- Conductivity using cellulose ester filter and treated pad for Fluorides, aerosol and gas	8/94
NIOSH 7905	Gas Chromatography- Phosphorus FPD using solid sorbent tube (Tenax GC) for Phosphorus	8/94

NIOSH 7904	Ion-Specific Electrode using PVC membrane filter and bubbler for Cyanides, aerosol and gas	8/94
NIOSH 7903	Ion Chromatography using solid sorbent tube (silica gel) for Inorganic Acids	8/94
NIOSH 7902	Ion-Specific Electrode using cellulose ester filter and treated pad for Fluorides, aerosol and gas	8/94
NIOSH 7901	Atomic Absorption- Graphite Furnace using Na_2CO_3-impregnated cellulose ester filter for Arsenic Trioxide, as As	8/94
NIOSH 7900	Atomic Absorption- Flame Arsine Generation using cellulose ester filter for Arsenic and Compounds, as As (except AsH_3 and AS_2O_3)	8/94
NIOSH 7702	X-Ray Fluorescence (XRF) Portable L-Shell Excitation using mixed cellulose ester filter for Lead	1/98
NIOSH 7701	Portable Anodic Stripping Voltammetry using mixed cellulose ester filter for Lead by Ultrasound	1/98
NIOSH 7700	Chemical Spot Test Kit using cellulose ester filter for Lead in Air	5/96
NIOSH 7402	Microscopy- Transmission Electron using cellulose ester filter for Asbestos by TEM	8/94
NIOSH 7400	Light Microscopy- Phase Contrast using cellulose ester filter for Asbestos and Other Fibers by PCM	8/94
NIOSH 7300	Inductively Coupled Argon Plasma- Atomic Emission Spectroscopy using cellulose ester filter for Elements by ICP	8/94
NIOSH 7105	Atomic Absorption Spectrophotometer, Graphite Furnace using cellulose ester filter for Lead	8/94
NIOSH 7102	Atomic Absorption- Graphite Furnace using cellulose ester filter for Beryllium and compounds, as Be	8/94
NIOSH 7082	Atomic Absorption Spectrophotometer, Flame using cellulose ester filter for Lead	8/94
NIOSH 7048	Atomic Absorption- Flame using cellulose ester filter for Cadmium and compounds, as Cd	8/94
NIOSH 7027	Atomic Absorption- Flame using cellulose ester filter for Cobalt and compounds, as Co	8/94
NIOSH 7024	Atomic Absorption- Flame using cellulose ester filter for Chromium and compounds, as Cr	8/94
NIOSH 6011	Ion Chromatography- Conductivity Detection using prefilter (PTFE) and silver membrane filter for Chlorine and Bromine	8/94

TABLE A3.1 (CONTINUED)
Results of the Survey of Ambient Air Measurement Methods for the 188 HAPs

NIOSH 6010	Spectrophotometry, Visible Absorption using solid sorbent tube (soda lime) for Hydrogen Cyanide	8/94
NIOSH 6009	Atomic Absorption- Cold Vapor using solid sorbent tube (Hopcalite) for Mercury	8/94
NIOSH 6002	UV-VIS Spectrometer using sorbent tube (coated silica gel) for Phosphine	1/98
NIOSH 6001	Atomic Absorption- Graphite Furnace using solid sorbent tube (charcoal) for Arsine	8/94
NIOSH 5700	HPLC- UV Detection using inhalable dust sampler with PVC filter for Formaldehyde on Dust (Textile or Wood)	8/94
NIOSH 5600	GC- Flame Photometric Detection (FPD) using quartz filter and solid sorbent tube (XAD-2) for Organophosphorus Pesticides	8/93
NIOSH 5523	Gas Chromatography- FID using XAD-7 OVS tube for Glycols	5/96
NIOSH 5522	HPLC- Fluorescence Detector/Electrochemical Detector using tryptamine/DMSO impinger for Isocyanates	1/98
NIOSH 5521	HPLC- Electrochemical and UV Detection using impinger for Monomeric Isocyanates	8/94
NIOSH 5517	Gas Chromatography- ^{63}Ni ECD using PTFE filter and solid sorbent tube (XAD-2) for Polychlorobenzenes	8/94
NIOSH 5516	HPLC- UV Detection using impinger for 2,4- and 2,6- Toluenediamine (in the presence of isocyanates)	8/94
NIOSH 5512	HPLC- UV Detection using mixed cellulose ester filter and bubbler for Pentachlorophenol	8/94
NIOSH 5510	Gas Chromatography- Electron Capture Detector (GC/ECD) using cellulose ester filter and solid sorbent tube (Chromosorb 102) for Chlordane	8/94
NIOSH 5509	HPLC- UV Detector using glass fiber filter for Benzidine	8/94
NIOSH 5506	HPLC- Fluorescence/UV Detection using PTFE filter and sorbent tube (XAD-2) for Polynuclear Aromatic Hydrocarbons	
NIOSH 5503	Gas Chromatography- ^{63}Ni ECD using glass fiber filter and solid sorbent (Florisil) for Polychlorobiphenyls	8/94

NIOSH Method	Description	Date
NIOSH 5502	Gas Chromatography- Electrolytic Conductivity Detector using glass fiber filter and bubbler for Lindane and Aldrin	8/94
NIOSH 5029	HPLC- UV and Electrochemical Detection using acid-treated glass fiber filter for 4,4'-Methylenedianiline	8/94
NIOSH 5020	Gas Chromatography- FID using cellulose ester filter for Bibutyl Phthalate	8/94
NIOSH 5011	Visible Absorption Spectrophotometry using PVC or cellulose ester filter for Ethylene Thiourea	8/94
NIOSH 5006	Visible Absorption Spectrophotometry using glass fiber filter for Carbaryl	8/94
NIOSH 5004	HPLC- UV Detection using cellulose ester filter for Hydroquinone	8/94
NIOSH 5001	HPLC- UV Detector using glass fiber filter for 2,4-D ((2,4-Dichlorophenoxy)acetic acid)	8/94
NIOSH 4000	Gas Chromatography- FID using diffusive sampler (activated carbon) for Toluene	8/94
NIOSH 3702	Gas Chromatography (Portable)- Photoionization Detector using ambient air or bag sample for Ethylene Oxide	8/94
NIOSH 3701	Gas Chromatography (Portable)- Photoionization Detector using Tedlar air bag for Trichloroethylene	8/94
NIOSH 3700	Gas Chromatography (Portable)- Photoionization Detector using Tedlar air bag for Benzene	8/94
NIOSH 3515	Visible Spectrophotometry using HCl bubbler for 1,1-Dimethylhydrazine	8/94
NIOSH 3514	HPLC- UV Detection using bubbler (Folin'd Reagent) for Ethylenimine	8/94
NIOSH 3512	HPLC- UV Detection using distilled water bubbler for Maleic Anhydride	8/94
NIOSH 3509	Ion Chromatography- Ion Pairing [2,3] using impinger for Aminoethanol compounds II	8/94
NIOSH 3507	HPLC- UV Detector using liquid in bubbler for Acetaldehyde	8/93
NIOSH 3503	Spectrophotometry, Visible Absorption using HCl bubbler for Hydrazine	8/94
NIOSH 3500	Visible Absorption Spectrometry using PTFE filter and sodium bisulfate impingers for Formaldehyde	8/94

TABLE A3.1 (CONTINUED)
Results of the Survey of Ambient Air Measurement Methods for the 188 HAPs

NIOSH 2549	Thermal Desorption Gas Chromatography- Mass Spectrometry using thermal desorption tube for Volatile Organic Compounds (Screening)	5/96
NIOSH 2546	Gas Chromatography- FID using solid sorbent tube (XAD-7) for Cresol (all isomers) and Phenol	8/94
NIOSH 2543	Gas Chromatography- ECD using solid sorbent tube (XAD-2) for Hexachlorobutadiene	8/94
NIOSH 2541	Gas Chromatography- FID using solid sorbent tube (XAD-2) for Formaldehyde	8/94
NIOSH 2539	Gas Chromatography- FID & GC/MS using solid sorbent tube (XAD-2) for Aldehydes, Screening	8/94
NIOSH 2538-1	Gas Chromatography-FID using solid sorbent tube (XAD-2) for Acetaldehyde	8/93
NIOSH 2537	Gas Chromatography- FID using solid sorbent tube (XAD-2) for Methyl Methacrylate	8/94
NIOSH 2535	HPLC- UV Detection using tube with reagent coated glass wool for Toluene 2,4-Diisocyanate	8/94
NIOSH 2530	Gas Chromatography- FID using solid sorbent tube (Tenax GC) for Diphenyl	8/94
NIOSH 2528	Gas Chromatography- FID using solid sorbent tube (Chromosorb) for 2-Nitropropane	8/94
NIOSH 2524	Gas Chromatography- Electrolytic Conductivity Detector (Sulfur Mode) using solid sorbent tube (Poropak P) for Dimethyl Sulfate	8/94
NIOSH 2522	Gas Chromatography (TEA) using solid sorbent tube (Thermosorb/N) for Nitrosamines	8/94
NIOSH 2520	Gas Chromatography- FID using solid sorbent tubes (charcoal) for Methyl Bromide	8/93
NIOSH 2519	Gas Chromatography- FID using solid sorbent tube (charcoal) for Ethyl Chloride	8/94
NIOSH 2518	Gas Chromatography- ^{63}Ni ECD using solid sorbent tube (Porapak T) for Hexachloro-1,3-cyclopentadiene	8/94
NIOSH 2515	Gas Chromatography- FID using solid sorbent tube (XAD-2) for Diazomethane	8/94

NIOSH 2514	HPLC- UV Detector using solid sorbent tube (XAD-2) for Anisidine; UV Detection using bubbler for Ethylenimine	8/94
NIOSH 2508	Gas Chromatography- FID using solid sorbent tube (charcoal) for Isophorone	8/94
NIOSH 2501	Gas Chromatography- Nitrogen specific detector using solid sorbent tube (XAD-2) for Acrolein	8/94
NIOSH 2500	Gas Chromatography- FID using solid sorbent tube (carbon molecular sieve) for Methyl Ethyl Ketone	8/94
NIOSH 2017	Gas Chromatography- FID using filter and solid sorbent tube (silica gel) for Aniline, o-Toluidine, and Nitrobenzene	1/98
NIOSH 2016	HPLC- UV Detection using silica gel cartridge for Formaldehyde	1/98
NIOSH 2008	Ion Chromatography- Conductivity Detection using solid sorbent tube (silica gel) for Chloroacetic Acid	8/94
NIOSH 2005	Gas Chromatography- FID using solid sorbent tube (silica gel) for Nitroaromatic compounds	1/98
NIOSH 2004	Gas Chromatography- FID using solid sorbent tube (silica gel) for Dimethylacetamide	8/94
NIOSH 2002	Gas Chromatography- FID using solid sorbent tube (silica gel) for Aromatic Amines	8/94
NIOSH 2000	Gas Chromatography- FID using solid sorbent tube (silica gel) for Methanol	1/98
NIOSH 1615	Gas Chromatography- FID using solid sorbent tubes (charcoal) for Methyl tert-Butyl Ether	8/94
NIOSH 1614	Gas Chromatography- ECD using solid sorbent tube (HBr-coated charcoal) for Ethylene Oxide	8/94
NIOSH 1612	Gas Chromatography- FID using solid sorbent tube (charcoal) for Propylene Oxide	8/94
NIOSH 1606	Gas Chromatography- FID using solid sorbent tube (charcoal) for Acetonitrile	1/98
NIOSH 1604	Gas Chromatography- FID using solid sorbent tube (charcoal) for Acrylonitrile	8/94
NIOSH 1602	Gas Chromatography- FID using solid sorbent tube (charcoal) for Dioxane	8/94
NIOSH 1600	Gas Chromatography- Sulfur FPD using solid sorbent tube (charcoal) and drying tube for Carbon Disulfide	8/94
NIOSH 1501	Gas Chromatography- FID using solid sorbent tube (charcoal) for Aromatic Hydrocarbons	8/94

TABLE A3.1 (CONTINUED)
Results of the Survey of Ambient Air Measurement Methods for the 188 HAPs

NIOSH 1500	Gas Chromatography- FID using solid sorbent tube (charcoal) for Hydrocarbons, 36-126 EC BP	8/94
NIOSH 1453	Gas Chromatography- FID using solid sorbent tube (carbon molecular sieve) for Vinyl Acetate	1/98
NIOSH 1450	Gas Chromatography- FID using solid sorbent tube (charcoal) for Esters I	8/94
NIOSH 1300	Gas Chromatography- FID using solid sorbent tube (charcoal) for Ketones I	8/94
NIOSH 1024	Gas Chromatography- FID using solid sorbent tube (charcoal) for 1,3-Butadiene	8/93
NIOSH 1019	Gas Chromatography- FID using solid sorbent tube (charcoal) for 1,1,2,2-Tetrachloroethane	8/94
NIOSH 1015	Gas Chromatography- FID using solid sorbent tube (charcoal) for Vinylidene Chloride	8/94
NIOSH 1014	Gas Chromatography- FID using solid sorbent tube (charcoal) for Methyl Iodide	8/94
NIOSH 1013	Gas Chromatography- Electrolytic Conductivity Detector (Hall) using solid sorbent tube (charcoal) for Propylene Dichloride	8/94
NIOSH 1010	Gas Chromatography- FID using solid sorbent tube (charcoal) for Epichlorohydrin	8/94
NIOSH 1009	Gas Chromatography- FID using solid sorbent tube (charcoal) for Vinyl Bromide	8/94
NIOSH 1008	Gas Chromatography- ^{63}Ni ECD using solid sorbent tube (charcoal) for Ethylene Dibromide	8/94
NIOSH 1007	Gas Chromatography- FID using solid sorbent tubes (charcoal) for Vinyl Chloride	8/94
NIOSH 1005	Gas Chromatography- FID using solid sorbent tubes (charcoal) for Methylene Chloride	1/98
NIOSH 1004	Gas Chromatography- FID using solid sorbent tube (charcoal) for Dichloroethyl Ether	8/94
NIOSH 1003	Gas Chromatography- FID using solid sorbent tube (charcoal) for Halogenated Hydrocarbons	8/94

NIOSH 1002	Gas Chromatography- FID using solid sorbent tube (charcoal) for Chloroprene	8/94
NIOSH 1001	Gas Chromatography- FID using solid sorbent tubes (charcoal) for Methyl Chloride	8/94
NIOSH 1000	Gas Chromatography- FID using solid sorbent tube (charcoal) for Allyl Chloride	8/94
EPA IO-Compendium Methods		
IO-5	Sampling and Analysis for Atmospheric Mercury	1/97
IO-3	Chemical Species Analysis of Filter Collected SPM	1/97
R-1	Collection of whole air in canisters, separation of co-collected water using a two-stage sorbent trap, thermal desorption, and analysis by GC with ion trap MS detection. Kelly, T.J. et al., Method development and field measurements for polar volatile organic compounds in ambient air, *Environ. Sci. Technol.*, 27(6), 1146, 1993.	
R-2	Sampling at 10 LPM using a Teflon microfiber matrix (4-in.-dia. filter) impregnated with 5 μm particles of AG-1 anion exchange resin; analyzed using BSTFA derivatization and EI-GC/MS. Burkholder, H. et al., Anion exchange resins for collection of phenols from air and water, in *Proc. 16th Ann. EPA Conf. Analysis of Pollutants in the Environment*, Norfolk, VA, May 1993.	
	Nishioka, M.G., Burkholder, H.M., Evaluation of an anion exchange resin for sampling ambient level phenolic compounds, Final Report for EPA Contract No. 68-02-4127/WA-69 and -80, Battelle, Columbus, OH, April 1990.	
R-3	Collection using three-stage tubes packed with Carbosieve S-III, Carbotrap, and Carbotrap C; desorption and refocusing onto an electrically cooled Carbosieve-III and Carbotrap sorbent bed; analysis by GC/FID and ECD.	
R-4	Pollack, A.J, Gordon, S.M. and Moschandreas, D.J., Evaluation of portable multisorbent air samplers for use with an automated multitube analyzer, Final Report for EPA Contract No.. 68-D0-0007/WA-27, September 1992. *Methods of Air Sampling and Analysis*, 3rd ed., Lodge, J.P. Jr., Ed., Intersociety Committee on Methods of Air Sampling and Analysis, Lewis Publishers, Inc., Chelsea, Michigan, 1989.	
[14]	Infrared Absorption Spectroscopy (using Saran or Mylar plastic bag sampler or silica gel) for VOCs	

TABLE A3.1 (CONTINUED)
Results of the Survey of Ambient Air Measurement Methods for the 188 HAPs

[205]   Determination of Fluoride Content of Plant Tissues (Potentiometric Method)

[301]   Determination of Particulate Antimony Content in the Atmosphere (using membrane, cellulose or glass fiber filters and visible absorption spectrophotometry)

[302]   Determination of Arsenic Content of Atmospheric Particulate Matter (using membrane or glass fiber filters and visible absorption spectrophotometry)

[317]   Determination of Elemental Mercury in Ambient and Workroom Air by Collection on Silver Wool and Atomic Absorption Spectroscopy

[606A]   Estimation of Airborne Radon-222 by Filter Paper Collection and Alpha Activity Measurements of its Daughters (Thomas Method or Modified Kusnetz Method)

[606B]   Determination of Airborne Radon-222 by its Absorption from the Atmospheric and Gamma Measurement (using charcoal adsorbent)

[804]   As, Se, and Sb in Urine and Air by Hydride Generation and Atomic Absorption Spectrometry (using cellulose acetate membrane filter)

[805]   Determination of Chlorine in Air (using midget impinger with sodium acetate and potentiometric analysis)

[809]   Determination of Fluorides and Hydrogen Fluoride in Air (using impingers with sodium hydroxide)

[822]   General Atomic Absorption Procedure for Trace Metals in Airborne Material Collected on Filters (using membrane filters) for particulate inorganics

[822B]   X-Ray Fluorescence Spectrometry for Multielemental Analysis of Airborne Particulate and Biological Material

[829]   Determination of Chloromethyl Methyl Ether (CMME) and Bis-Chloromethyl Ether (Bis-CME) in Air (using GC-ECD and impingers with a methanolic solution of the sodium salt of 2,4,6-trichlorophenol)

[831]   Determination of p,p-Diphenylmethane Diisocyanate (MDI) in Air (using midget impingers with hydrochloric and acetic acids and visible absorption spectrophotometry)

[835]   Determination of EPN, Malathion and Parathion in Air (using glass fiber filters and GC-flame photometric detection)

[836]   Determination of Total Particulate Hydrocarbons (TpAH) in Air: Ultrasonic Extraction Method (using glass fiber filters and HPLC-UV)

[837]   Determination of 2,4-Toluene Diisocyanate (TDI) in Air (using midget impingers with hydrochloric and acetic acids and visible absorption spectrophotometry)

R-5   High volume air sampling with glass fiber filter and polyurethane foam sorbent; solvent extraction and chromatographic cleanup; analysis by high resolution gas chromatography and high resolution mass spectrometry (HRGC/HRMS), with multiple isotopically labelled internal surrogate standards. Analysis based on guidelines of EPA Methods 8280 and 8290.

Hunt, G.T. and Maisel, B.E., Atmospheric concentrations of PCDDs/PCDFs in southern California, *J. Air Waste Mgt. Assoc.*, 42, 672, 1992.

R-6   Automated gas chromatography with detection by ECD and FID. Sample collection performed hourly using a three-stage sorbent trap, with refocusing on a cryogenic (-186 °C) trap for analysis.

Purdue, L.J. et al., Atlanta Ozone Precursor Monitoring Study Data Report, EPA/600/R-92/157, U.S. EPA, Washington, D.C., September 1992.

R-7   Canister collection of whole air samples; analysis by gas chromatography/multiple detector (ECD, FID, PID) method.

McAllister, R. et al., 1990 Urban Air Toxics Monitoring Program. Report No. EPA-450/4-91-024, prepared for U.S. EPA by Radian Corporation, Research Triangle Park, NC, June 1991.

R-8   Sampling with denuder/filter/XAD resin combinations, extraction, and analysis by ion chromatography.

Eatough, D.J. et al., Identification of gas-phase dimethyl sulfate and monomethyl hydrogen sulfate in the Los Angeles Atmosphere, *Environ. Sci. Technol.*, 20, 867, 1986.

Hansen, L.D., White, V.F. and Eatough, D.J., Determination of gas-phase dimethyl sulfate and monomethyl hydrogen sulfate, *Environ. Sci. Technol.*, 20, 872, 1986.

R-9   Four distinct methods: 1) collection on Thermosorb A, solvent extraction, and analysis by GC with nitrogen-selective detector; 2) collection on Tenax, thermal desorption, and GC/MS analysis; 3) grab sampling with analysis by portable GC/FID; 4) atmospheric pressure ionization quadrupole MS.

Clay, P.F. (NUS Corp., Bedford, Mass), Spittler, T.M. (U.S. EPA, Region I, Lexington, Mass), Determination of airborne volatile nitrogen compounds using four independent techniques, *Proc. Natl. Conf. Manage. Uncontrolled Hazard Waste Sites. Hazard. Mater. Control Res. Inst.*: Silver Spring, MD, 100, 1983.

R-10   Cryogenic trapping, thermal desorption, and GC analysis with flame photometric detection.

TABLE A3.1 (CONTINUED)
Results of the Survey of Ambient Air Measurement Methods for the 188 HAPs

Maroulis, P.J., Torres, A.L. and Bandy, A.R., Atmospheric concentrations of carbonyl sulfide in the southwestern and eastern United States, *Geophys. Res. Lett.*, 4, 510, 1977.

Torres, A.L. et al., Atmospheric OCS measurements on Project Gametag, *J. Geophys. Res.*, C12, 7357, 1980.

R-11   Cryogenic trapping, thermal desorption, and sequential GC analysis of two samples collected simultaneously with flame photometric detection.

Maroulis, P.J. and Bandy, A.R., Measurements of atmospheric concentrations of CS2 in the eastern United States, *Geophys. Res. Lett.*, 7, 681, 1980.

R-12   Collection on Tenax sorbent, thermal desorption to a cryogenic focussing trap, and GC/MS analysis.

Pellizzari, E.D. and Bunch, J.E., Ambient air carcinogenic vapors: improved sampling and analytical techniques and field studies, EPA-600/2-79-081, NTIS No. PB-297-932, U.S. EPA, Research Triangle Park, NC, May 1979.

R-13   Collection in canisters or cryogenically, with four detection methods. (1) Ion Trap GC/MS–detection limit of 1 pptv. (2) Quadrupole GC/MS with Selective Ion Monitoring–detection limit of 10 pptv. (3) Gas Chromatography with Photoionization Detection–detection limit 10 pptv. (4) Quadrupole GC/MS with Full Scan Monitoring–detection limit of 0.1 ppbv. Also Portable Gas Chromatograph with Photoionization Detection–detection limit of 0.1 ppbv.

Havlicek, S.C. et al., Assessment of ethylene oxide concentrations and emissions from sterilization and fumigation processes (PB93-216793; available NTIS). Final report by Coast-to-Coast Analytical Services, Inc., San Luis Obispo, CA, to California Air Resources Board, Sacramento, CA, Contract No. ARB-A832-125, 78 pp, May 1992.

R-14   Collection on XAD-4 resin Soxhlet extraction in dichloromethane and then in ethylacetate, analysis by GC/MS with positive chemical ionization.

Chuang, J.C. et al., Polycyclic aromatic hydrocarbons and their derivatives in indoor and outdoor air in an eight-home study, *Atmos. Envir.*, 25(3), 369, 1991. R-15 Collection on XAD-2 or PUF, Soxhlet extraction in 10 percent ether/hexane or methylene chloride, analysis by GC/MS.

Chuang, J.C., Hannan, S.W. and Wilson, N.K., Field comparison of polyurethane foam and XAD-2 resin for air sampling for polynuclear aromatic hydrocarbons, *Envir. Sci. Tech.*, 21(8), 798, 1987.

R-16   Air sampling methods discussed with references; minimum detectable levels provided for a number of particulate and gaseous radionuclides in air presented, assuming standard gamma-scan–400 to 512 multichannel analyzer–4 x 4-inch NaI (Te) detector or liquid scintillation counting, with references.

CRC Handbook of Environmental Radiation, Ed. Alfred W. Clement, Jr., CRC Press, Inc., Boca Raton, FL, 1982.

R-17   Background information on radioactivity detectors, measurement procedures, quality assurance, and statistical analysis of radioactivity measurements.

Handbook of Radioactivity Measurements Procedures, NCRP Report No. 58, National Council on Radiation Protection and Measurements, Bethesda, MD, 1985.

R-18   Collection on Teflon filter and iodated activated carbon, acid digestion, and analysis by cold vapor atomic absorption.

Lindbergh, S.E. et al., Atmospheric concentrations and deposition of Hg to a deciduous forest at Walker Branch Watershed, Tennessee, USA, Water, Air, Soil Poll., 56, 577, 1991.

Turner, R.R. et al., Mercury in ambient air at the Oak Ridge Y-12 plant July 1986 through December 1990, Y-12 report Y/TS-574, Oak Ridge National Laboratory, Oak Ridge, Tennessee, 1991.

R-19   Collection on alkaline-impregnated glass fiber filters, aqueous extraction, and ion chromatographic analysis using "negative" UV photometric detection with a strongly UV-absorbing eluent.

Grosjean, D., Liquid chromatography analysis of chloride and nitrate with "negative" ultraviolet detection: ambient levels and relative abundance of gas-phase inorganic and organic acids in southern California, *Environ. Sci. Technol.*, 24, 77, 1990.

R-20   Revision of ASTM Method D-3266, involving collection of HF on alkaline-impregnated tape, aqueous extraction, and analysis by ion-selective electrode.

Zankel, K.L. et al., Measurement of ground-level concentrations of hydrogen fluoride, *J. Air Poll. Control Assoc.*, 37, 1191, 1987.

R-21   Collection on Nuclepore (i.e., polycarbonate) filters, carbon coating by vapor deposition, and electron microscopic examination.

Samudra, A.U., Harwood, C.F. and Stockhalm, J.D., Electron microscope measurement of airborne asbestos concentration. EPA-600/1-77-178, U.S. EPA, Research Triangle Park, NC, 1977.

See also discussion in: *Asbestiform Fibers: Nonoccupational Health Risks*, published by National Academy Press, National Academy of Sciences, Washington, D.C., pp 82-96, 1984.

TABLE A3.1 (CONTINUED)
Results of the Survey of Ambient Air Measurement Methods for the 188 HAPs

R-22   Sampling using a chilled acetone collection medium to trap hydrazines and convert to stable derivatives; acetone solution analyzed directly for derivatives using a gas chromatograph with nitrogen-specific detectors.

R-23   Collection and derivatization in toluene solution containing N-(4-nitrobenzyl)-N-n-propylamine hydrochloride (NBPA); analysis by HPLC and UV detection.

Holdren, M.W., Spicer, C.W. and Riggin, R.M., Gas phase reaction of toluene diisocyanate with water vapor, *Am. Ind. Hygiene Assoc. J.*, 45, 626, 1984.

Dunlap, K.L., Sandridge, R.L. and Keller, J., Determination of isocyanates in working atmospheres by high-speed liquid chromatography, *Anal. Chem.*, 45, 497, 1976.

R-24   Continuous real-time monitoring based on color formation in a substrate-impregnated tape, with electro-optical measurement of color intensity. The unit has a useful range up to 200 ppbv of TDI, with a detection limit of 1 ppbv. The monitor is sold as the TLD-1 by MDA Scientific, Inc., Lincolnshire, Illinois.

R-25   Sampling at 0.1 LPM using 200–400 mesh granular AG-1 anion exchange resin; analyzed using EI-GC/MS and/or GC/FID; methylation and GC/ECD for chlorinated phenols.

Nishioka, M.G. and Burkholder, H.M., Evaluation of an anion exchange resin for sampling ambient level phenolic compounds, Final Report for EPA Contract No. 68-02-4127/WA-69 and -80, Battelle, Columbus, OH, April 1990.

R-26   Passive sampling using commercial samplers with activated carbon as sorbent; solvent extraction and concentration; analysis by GC/MS.

Shields, H.C. and Weschler, C.J., Analysis of ambient concentrations of organic vapors with a passive sampler, *J. Air. Poll. Control Assoc.*, 37, 1039, 1987.

R-27   Collection on polyurethane foam (sampling for 24 h at 3.8 L/min), solvent extraction and evaporative concentration, analysis by GC/ECD and GC/MS (multiple ion mode).

Immerman, F.W. and Schaum, J.L., Final Report of the Nonoccupational Pesticide Exposure Study (NOPES), EPA-600/3-90-003, U.S. EPA, Research Triangle Park, NC, 1990.

Whitmore, R.W. et al., Nonoccupational exposures to pesticides for residents of two U.S. cities, *Arch. Environ. Contam. Toxicol.*, 26, 47, 1994.

R-28 Collection on glass fiber filters and polyurethane foam or Florisil sorbents. Solvent extraction, cleanup on Florisil, evaporative concentration, and analysis by GC/ECD.

Atlas, E. and Giam, C.S., Ambient concentration and precipitation scavenging of atmospheric organic pollutants, *Water Air Soil Poll.*, 38, 19, 1988.

Chang, L.W., Atlas, E. and Giam, C.S., Chromatographic separation and analysis of chlorinated hydrocarbons and phthalic acid esters from ambient air samples, *Int. J. Environ. Anal. Chem*, 19, 145. 1985.

R-29 Collection with glass fiber filters and polyurethane foam; solvent extraction, cleanup on Florisil, evaporative concentration; analysis by GC/ECD and GC/MS.

Hoff, R.M., Muir, D.C.G. and Grift, N.P., Annual cycle of polychlorinated biphenyls and organohalogen pesticides in air in southern Ontario 1. Air concentration data, *Environ. Sci. Technol*, 26, 266, 1992.

R-30 High- or low-volume sampling of ambient air. Collection on glass fiber filters and polyurethane foam, solvent extraction, evaporative concentration, and analysis by GC/ECD or GC/MS.

Lewis, R G. et al., Measurement of atmospheric concentrations of common household pesticides: a pilot study, *Environ. Monitoring and Assess.*, 10, 59, 1988. R-31 Collection on glass fiber filters and on any of three sorbents: polyurethane foam, XAD-2 resin, or Tenax GC. Solvent extraction, cleanup, and analysis by GC/ECD.

Billings, W.N. and Bidleman, T.F., High-volume collection of chlorinated hydrocarbons in urban air using three solid sorbents, *Atmos. Environ.*, 17, 383, 1983.

R-32 Filter sampling at 1.5 LPM, total sampling volume of 240-L using PVC filters; extraction using buffer solution, pH = 4.5; trisodium pentacyanoaminoferrate reagent; colorimetric analysis in 30 min at 475 nm in 1-cm glass cells; paper provides sampling conditions, extraction solvents, reagents, analytical conditions, and detection ranges for colorimetric determination.

Freixa, A. and Magti, A., Application of colorimetric techniques to the measurement of air pesticide content, Pergamon Ser. *Environ. Sci.*, 7, 297, 1982.

R-33 Gas chromatography with flame photometric detection (GC-FPD) using a column (0.5 m x 2.5 mm i.d.) packed with GDX-101. Phosphine retention time 19 s at 80 C; ratio of H to 0 of 10:3. Qi, Xiaofei, Quantitative determination of trace phosphine in ambient air by gas chromatography with a flame photometric detector, Sepn, 5(4): 243-5 (in Chinese), 1987.

R-34 Chemiluminescence emission from arsine due to reaction of sampled air with ozone.

Inone, K., Suzuki, M. and Kawabayashi, O., Method and apparatus for chemiluminescence analyses, Ger. Offen., DE 3525700/A1/860206 (German patent). (reports detection limit of 1 ppb for arsine), 1986.

TABLE A3.1 (CONTINUED)
Results of the Survey of Ambient Air Measurement Methods for the 188 HAPs

Fraser, M.E., Stedman, D.H. and Henderson, M.J., Gas-phase chemiluminescence of arsine mixed with ozone, *Anal. Chem.*, 54(7), 1200-1, 1982.

R-35   Analysis using gas chromatography and an alkali flame ionization detector (N-detector); acidified aqueous solutions directly injected on the column.
Donike, M., Gas chromatographic trace analysis of hydrocyanic acid in the nano- and picogram range, Mitteilungsbl. GDCh-Fachgrappe Lebensmittelchem. Gerichtl. Chem., 1974, 28(1-2): 46-52 (in German).

R-36   Collection of particulate material from air, analysis by HPLC with electrochemical detection.
Riggin, R.M. et al., Determination of benzidine, related congeners, and pigments in atmospheric particulate matter, *J. Chromatogra. Sci.*, 21, 321, 1983.

R-37   Derivatization of amines to the corresponding amides by reaction with a perfluoro-acid anhydride, gas chromatographic separation, and analysis by N-selective thermionic detection.
Skarping, G., Renman, L. and Dalene, M., Trace analysis of amines and isocyanates using glass capillary gas chromatography and selective detection. II. Determination of aromatic amines as perfluorofatty acid amides using nitrogen-selective detection, *J. Chromatogr.*, 270, 207, 1983.

R-38   GC/ECD method for 2,4-D salts and acid.
Nishioka, M. et al., Simulation of track-in of lawn-applied herbicide acids from turf to home: Comparison of dislodgeable turf residues with carpet dust and carpet surface residues, prepared for submission to *Environ. Sci. Technol.*, 1993.

R-39   Collection on Tenax-GC sorbent, thermal desorption, gas chromatography with Hall electrolytic chlorine-sensitive detection.
Matienzo, L.J. and Hensler, C.J., Determination of N,N-dimethylcarbamoyl chloride (DMCC) in air, *Am. Indus. Hygiene Assoc. J.*, 43, 838, 1982.

R-40   Collection with glass fiber filter and Tenax GC sorbent, thermal desorption, cryogenic concentration, and analysis by GC/MS.
Krost, K.J. et al., Collection and analysis of hazardous organic emissions, *Anal. Chem.*, 54, 810, 1982.

R-41   Collection in methylisobutyl ketone, gas chromatography with sulfur-selective detection. Alternatively, collection and derivatization on the pre-coated walls of a diffusion denuder tube, and determination by HPLC with UV detection.
Oldeweme, J. and Klockow, D., Chromatographic procedures for the determination of 1,3-propanesultone in workplace air, Fresenius Z. *Anal. Chem.*, 325, 57, 1986.

R-42   Collection in aqueous KOH solution containing methanol and hydroxyl amine, to form a derivative. The iron complex of that derivative is determined quantitatively by absorbance of 530 nm.

Jozwicka, J., Spectrophotometric method for determination of monochloroacetic acid vapors in workplace air, *Wlokna Chem.*, 16, 394, 1990.

R-43   Workplace air monitoring of caprolactam; collection on filter and XAD-2 tubes or XAD-2 tubes only; desorption with methanol containing 2% water, or with MeCN; analysis by GC or HPLC. Sampled air volume of 100 L yields detection limit of 0.20 mg/m3 using HPLC analysis, and 0.10 mg/m3 using GC analysis.

Nau, D.R. et al., Validation study of a method for monitoring personnel exposure to caprolactam, *Proc Symp. Ind. Approach Chem. Risk Assess.: Caprolactam Relat. Compd. Case Study*, 275-91. Ind. Health Found.: Pittsburgh, PA, 1984.

R-44   Detection of caprolactam in workplace air and toxicol. studies; aerosols sampled on filter AFA-KhA-20 at 2 L/min, extracted with di-Et ether or a 1:1 EtOH/ether mixture; evaporated, and chromatographic drying in Cl, analysis by thin-layer chromatography with a mobile alcohol solvent system; and development with o-tolidine solution or fresh KI-starch reagent. Detection limit is 0.005 mg/m3; cyclohexanone, hydroxylamine, and NH3 stated not to interfere with method.

Ledovskikh, N.G., Sensitive method for the determination of caprolactam in air, Gig. r. Prof. Zabol., 10: 52-3 (in Russian), 1982.

R-45   Collection by charcoal adsorbent tube (or silica gel tube under high humidity conditions); desorption using distilled water then carbon disulfide; analysis of both layers by GC-FID.

Langhorst, M.L., Glycol ethers--validation procedures for tube/pump and dosimeter monitoring methods, *Am. Ind. Hyg. Assoc. J.*, 45, 416, 1984.

R-46   Personal air sampling through polyurethane foam plug; extraction in hexane; solvent transfer to toluene; analysis by GLC.

Nigg, H.N. and Stamper, J.H., exposure of spray applicators and mixer-loaders to chlorobenzilate miticide in Florida citrus groves, *Arch. Environ. Contam. Toxicol.*, 12, 477, 1983.

R-47   Headspace gas chromatography is applied for the analysis of water in liquid and solid samples with the preferred quantitation technique being standard addition.

Kolb, B. and Auer, M., Analysis for water in liquid and solid samples by headspace gas chromatography. Part I: liquid and soluble solid samples, *Fresenius. Z. Anal. Chem.*, 336, 291, 1990.

R-48   Sampling via an activated carbon fiber felt put between quartz filters and determination by gas chromatography-mass spectroscopy.

Suzuki, S., Simultaneous determination of airborne pesticides by GC/MS, *Bunseki Kagaku*, 41:115-24 (in Japanese), 1992.

R-49   Collection on XAD resins and determination by gas chromatography-mass spectroscopy and a nitrogen-phosphorous detector.

TABLE A3.1 (CONTINUED)
Results of the Survey of Ambient Air Measurement Methods for the 188 HAPs

Yeboah, P.O. and Kilgore, W.W., Analysis of airborne pesticides in a commercial pesticide storage building, *Bull. Environ. Contam. Toxicol.*, 32, 629, 1984.

R-50   High volume sampling with collection on a cartridge containing PUF/Tenax/PUF. Multiple extraction, derivation, and analysis by GC/MS or GC/ECD.

McConnell, L.L. et al., Development of a collection method for chlorophenolic compounds in air, in *Proc. 1989 EPA/AWMA Symp. Measurement of Toxic and Related Air Pollutants*, Publication VIP-13, EPA-600/9-89-060, Air and Waste Mgt. Assoc., Pittsburgh, 623, 1989.

R-51   Collection with glass fiber filters and polyurethane foam sorbent. Solvent extraction after spiking with 13C-labeled isomers, chromatographic cleanup of the extracts, and evaporative concentration. Analysis by GC/MS using electron impact or electron capture negative ion modes.

Eitzer, B.D. and Hites, R.A., Polychlorinated dibenzo-p-dioxins and dibenzofurans in the ambient atmosphere of Bloomington, IN, *Environ. Sci. Technol.*, 23, 1389, 1989.

Edgerton, S.A. et al., Ambient air concentrations of polychlorinated dibenzo-p-dioxins and dibenzofurans in Ohio: Sources and health risk assessment, *Chemosphere*, 18, 1713, 1989.

R-52   High-volume sampling with a Teflon-impregnated glass fiber filter and three PUF sorbent plugs in series; addition of deuterated internal standards; solvent extraction and evaporative concentration; analysis by HPLC with UV detection.

Arey, J. et al., Polycyclic aromatic hydrocarbon and nitroarene concentrations in ambient air during a wintertime high-NOx episode in the Los Angeles basin, *Atmos. Environ.*, 21, 1437, 1987.

R-53   Collection on Teflon-impregnated glass fiber filters and XAD-2 resin sorbent; multiple solvent extractions with addition of deuterated internal standards, and separation of acid and base/neutral fractions by HPLC. Analysis by negative chemical ionization GC/MS.

Nishioka, M.G. and Lewtas, J., Quantification of nitro- and hydroxylated nitro-aromatic/polycyclic aromatic hydrocarbons in selected ambient air daytime winter samples, *Atmos. Environ.*, 26A, 2077, 1992.

R-54   Sampling with glass fiber filters and Tenax-GC and polyurethane foam sorbent traps; solvent extraction with addition of deuterated internal standards; analysis by GC/MS with electron impact ionization.

Leuenberger, C., Ligocki, M.P. and Pankow, J.F., Trace organic compounds in rain. 4. Identities, concentrations, and scavenging mechanisms for phenols in urban air and rain, *Environ. Sci. Technol.*, 19, 1053, 1985.

R-55   Continuous monitoring in air using ion mobility mass spectrometry.
Leasure, C.S. and Eiceman, G.A., Continuous detection of hydrazine and monomethylhydrazine using ion mobility spectrometry, *Anal. Chem.*, 57, 1890, 1985.

R-56   Derivatization on-column with an alkali metal salt of 2,4,6-trichlorophenol to form a derivative, which is determined immediately by GC with electron capture detection.
Kallos, G.J., Albe, W.R. and Solomon, R.A., On-column reaction gas chromatography for determination of chloromethyl methyl ether at one part-per-billion level in ambient air, *Anal. Chem.*, 49, 1817, 1977.

R-57   Collection with glass fiber filters and Tenax-GC sorbent; solvent extraction, evaporative concentration, and analysis by GC/MS.
Cautreels, W. and van Cauwenberghe, K., Experiments on the distribution of organic pollutants between airborne particulate matter and the corresponding gas phase, *Atmos. Environ.*, 12, 1133, 1978.

R-58   Low volume collection on Tenax-GC, thermal desorption, cryofocusing, and GC/MS analysis in a mobile field sampling laboratory.
Haggert, B. and Havkov, R., Design and implementation of a mobile monitoring unit (MMU) to measure ambient volatile organic compounds, paper 84-17.2, presented at the 77th Annual Meeting, Air Pollution Control Association, San Francisco, CA, June 1984.

R-59   Method stated to be collection of whole air in sampling bags, with analysis by GC with photoionization detection.
Hunt, W.F., Jr., Faoro, R.B. and Freas, W., Report on the Interim Data Base for State and Local Air Toxic Volatile Organic Chemical Measurements, Report No. EPA-450/4-86-012, U.S. EPA, Research Triangle Park, NC, 1986.

R-60   Modified version of Compendium Method TO-8, using C18 Sep-Pak cartridges coated with NaOH for sampling, with analysis by HPLC. The resolution of the HPLC analysis is improved by changing the pH of the acetate buffer, and by using a sequential bonding end-capped column.
Bratton, S.A., Sampling and measurement of phenol and methylphenols (cresols) in air by HPLC using a modified Method TO-8, in *Measurement of Toxic and Related Air Pollutants, Proc. 1992 EPA/AWMA Int. Symp.*, EPA Report No. EPA-600/R-92/131, Publication VIP-25, Air and Waste Mgt. Assoc., Pittsburgh, PA, 719, 1992.

TABLE A3.1 (CONTINUED)
Results of the Survey of Ambient Air Measurement Methods for the 188 HAPs

R-61   Adsorption of pentachlorophenol (PCP) onto OV-17 stationary phase, with collection /thermal desorption on a 2-min cycle. Analysis of desorbed PCP by atmospheric pressure chemical ionization tandem mass spectrometry. Detection limit 40 ng/m3.

        DeBrou, G.B., Ng, A.C. and Karellas, N.S., Near real-time measurements of pentachlorophenol in ambient air by mobile mass spectrometry, in *Measurement of Toxic and Related Air Pollutants, Proc. 1992 EPA/AWMA Int. Symp.,*, EPA Report No. EPA-600/R-92/131, Publication VIP-25, Air and Waste Mgt. Assoc., Pittsburgh, PA, 838, 1992.

R-62   Collection of 2,4-toluene diisocyanate in a derivatizing solution of 1-(2-pyridyl)piperazine in toluene, in impingers. The stable TDI/urea derivative is determined by HPLC. Limit of detection for TDI is 116 ng/m3, limit of quantitation is 351 ng/m3.

        Wilshire, F.W. et al., Development and validation of a source test method for 2,4-toluene diisocyanate, in *Measurement of Toxic and Related Air Pollutants, Proc. 1993 EPA/AWMA Int. Symp.,*EPA Report No. EPA-600/A93/024, Publication VIP-34, Air and Waste Mgt. Assoc., Pittsburgh, PA, 399, 1993.

R-63   Collection of airborne asbestos on a polycarbonate or mixed cellulose ester filter, deposition of carbon under vacuum, and dissolution of the original filter material. Analysis and counting are conducted by analytical electron microscopy, with identification by electron diffraction and energy dispersive x-ray spectroscopy.

        Doorn, S.S. and Burris, S.B., Airborne asbestos analysis by analytical electron microscopy, in *Measurement of Toxic and Related Air Pollutants, Proc. 1992 EPA/AWMA Int. Symp.,*, EPA Report No. EPA/600/9-91/018, Publication VIP-21, Air and Waste Mgt. Assoc., Pittsburgh, PA, 226, 1991.

R-64   Cryogenic concentration of a 100 mL air sample, separation by two dimensional gas chromatography, with flame ionization detection.

        Fung, K., A method for the measurement of alcohols and MTBE in ambient air, in *Measurement of Toxic and Related Air Pollutants, Proc. 1992 EPA/AWMA Int. Symp.,*, EPA Report No. EPA/600/9-91/018, Publication VIP-21, Air and Waste Mgt. Assoc., Pittsburgh, PA, 770, 1991.

R-66   Collection using filter and sorbent; analysis using gas chromatography/high-resolution mass spectrometry (GC/HRMS).

R-65   Sorbent collection on Porapak Q; thermal desorption; dual-column GC analysis with FID. Frankel, L. S. and Black, R. F., Automatic gas chromatography monitor for the detection of part-per-billion levels of bis(chloromethyl) ether, *Analyt. Chem.*, 48, 732, 1976.

R-67   DeRoos, F.L. et al., Evaluation of the EPA high-volume air sampler for collection and retention of polychlorinated dibenzo-p-dioxins and polychlorinated dibenzofurans, in *Measurement of Toxic and Related Air Pollutants, Proc. 1986 EPA/APCA Int. Symp.*, EPA Report No. EPA/600/9-86/013, Publication VIP-7, Air Pollution Control Association, Pittsburgh, PA, pp.217-229 (1986); see also EPA Report No. EPA-600/4-86/037.

Collection using filter and sorbent; analysis using gas chromatography/high-resolution mass spectrometry (GC/HRMS).

R-68   Harless, R.L. et al., Determination of polychlorinated dibenzo-p-dioxins and dibenzofurans in stack gas emissions and ambient air, in *Measurement of Toxic and Related Air Pollutants, Proc. 1988 EPA/APCA Int Symp*, EPA Report No. EPA/600/9-88/015, Publication VIP-10, Air Pollution Control Association, Pittsburgh, PA, 613, 1988.

The report documents the measurement of airborne levels of two pesticides, carbofuran and captan. The sampling medium was XAD-4 resin. Analyses were carried out with a gas chromatograph equipped with a Hall electrolytic conductivity detector.

R-69   Shibamoto, T., Mourer, C. and Hall G., Pilot monitoring study of two pesticides in air, Final Report to California State Air Resources Board, Sacramento, Research Div., Report No. ARB/R-95/579, 1993.

The sampling method consisted of a combined filter and XAD-2 adsorbent bed. Analysis was accomplished with HPLC-UV detection. Fourteen organonitrogen pesticides were evaluated using NIOSH guidelines and procedures.

R-70   Kennedy, E.R. et al., A sampling and analytical method for the simultaneous determination of multiple organonitrogen pesticides in air, *Am. Ind. Hyg. Assoc. J.*, 58(1), 720, 1997.

A method is presented that details procedures used to enhance peak identification. Procedures include changing column types, column parameters, fractionating the sample, etc.

R-71   Lewis, R.G., Determination of pesticides and polychlorinated biphenyls in indoor air by gas chromatography, IARC Sci. Publ., 109 (Environmental Carcinogens, Methods of Anlysis, and Exposure Measurement, Vol. 12) 353-376 (1993).

Air samples were taken indoors after application of chemicals. Residues were low (ca. 0.1 $\mu g/m^3$) during the 30-day sampling period.

Leidy, R.B. and Stout, D.M. II, Residues of chlorpyrifos and dichlorvos indoors following a perimeter house application, Book of Abstracts, 211th ACS National Meeting, New Orleans, LA, AGRO-192, American Chemical Society, Washington, D.C., 1996.

R-72   Samples were collected on a combination cellulose ester membrane (MCEM) filter and a silica gel Sep-Pak adsorbent. Each was extracted with 1% acetic acid. HPLC with fluorescence detection was used. Method detection was 0.3 $\mu g/m^3$.

TABLE A3.1 (CONTINUED)
Results of the Survey of Ambient Air Measurement Methods for the 188 HAPs

R-73   Risner, C.H., The quantification of hydroquinone, catechol, phenol, 3-methylcatechol, scopoletin, m+p-cresol and o-cresol in indoor air samples by high-performance liquid chromatography, *J. Liq. Chromatogr.*, 16(18), 4117, 1993.

Derivatization with pentafluorobenzyl bromide followed by gas chromatography/ion trap mass spectrometry is used to determine oxidation products of biogenic emissions. Acrylic acid was also identified and quantified with this method.

Chien, C-J. et al., Analysis of airborne carboxylic acids and phenols as their pentafluorobenzyl derivatives: gas chromatography/Ion trap mass spectrometry with a novel chemical ionization reagent, PFBOH, *Environ, Sci, Technol.*, 32(2), 299, 1998.

R-74   Nine residences were monitored for crop-related pesticides and PAH compounds.

Mukerjee, S. et al., An environmental scoping study in the lower Rio Grande Valley of Texas, B Part III, Residential microenvironmental monitoring for air, house dust, and soil, *Environ. Int.*, 23(5), 657, 1997.

R-75   Organochlorine pesticides and polychlorinated biphenyls at pg/m^3 concentrations were determined using high volume air sampling techniques.

McConnell, L.L. et al., Air concentrations of organochlorine insecticides and polychlorinated biphenyls over Green Bay, WI, and the four lower Great Lakes, *Environ. Pollut.*, 101(3), 391, 1998. R76 Cartridges containing florisil and foam plugs were used to collect air. Supercritical fluid extraction was used to extract material. Gas chromatography with electron capture detection was used for analyses. Detection limits of 0.1 ng/m^3 were obtained.

Swami, K., Narang, A.S. and Narang, R.S., Determination of chlordane and chlorpyrifos in ambient air at low nanogram-per-cubic meter levels by supercritical fluid xxtraction. *J. AOAC Int.*, 80(1), 74, 1997.

R-77   A patent was filed for a procedure to measure phosphorus. Phosphorus is converted to phosphorus monoxide which is reacted with ozone to convert the monoxide to a dioxide. A light measuring device then measures the intensity of emitted light.

Stedman, D.H. and Meeks, P.A., Method to detect phosphorus, U.S., patent number 5702954A.

R-79   PCB and PAHs were measured at various locations. Daytime vs. nighttime and urban vs. non-urban results were compared.

Simcik, M.F. et al., Urban contamination of the Chicago/Coastal Lake Michigan atmosphere by PCBs and PAHs during AEOLOS, *Environ. Sci. Technol.*, 31(7), 2141, 1997.

R-80   Analytical method was developed to determine PAHs in house dust and soil. The purpose was concentration profiles of PAHs in house dust and track-in soil.

Chuang, J.C. et al., Monitoring methods for polycyclic aromatic hydrocarbons and their distribution in house dust and track-in soil, *Environ. Sci. Technol.*, 29(2), 494, 1995.

R-81   A sorbent-based gas chromatographic method provides near real time monitoring of toxic gases. The system was demonstrated on-site. Deactivation and passivation techniques were critical to optimize method performance.

Lattin, F.G. and Paul, D.G., A method for near real time continuous air monitoring of phosgene, hydrogen cyanide, and cyanogen chloride, *Proc. SPIE-Int. Soc. Opt. Eng.*, 2835 (*Advanced Technologies for Environmental Monitoring and Remediation*), 180, 1996.

R-82   XAD-7 is the adsorbent for the collection of each analyte. Methanol was used as the extracting solvent. The use of a Stabilwax-DA analytical column resulted in enhanced peak shape and lower detection limits.

Pendergrass, S.M., An alternative method for the analysis of phenol and o-, m-, and p-cresol by capillary GC/FID, *Am. Ind. Hyg. Assoc. J.*, 55(11), 1051, 1994.

R-83   The impinger method has a limit of detection of 25 ng per 40 mL of solution. The method compared well with two NIOSH methods.

Wyatt, J.R. et al., Coulometric method for the quantification of low-level concentrations of hydrazine and monomethylhydrazine, *Am. Ind. Hyg. Assoc. J.*, 54(6), 285, 1993.

R-84   Air is drawn through a paper tape treated with vanillin (4-hydroxy-3-methoxybenzaldehyde). The contaminated air reacts with vanillin to develop a yellow color. The density of the color is proportional to the concentration of hydrazine and methylhydrazine. Method detection is low ppb.

Young, R.C., McBrearty, C.F. and Curran, D.J., Active hydrazine vapor sampler (AHVS), NTIS Report N93-22149/7, 1993.

4 Concentrations of the 188 HAPs in Ambient Air

4.1 INTRODUCTION

As earlier chapters have made clear, the 188 HAPs are a diverse group that includes nonpolar and polar volatile organic compounds (VOCs), semivolatile organic compounds (SVOCs) (including pesticides and polycyclic aromatic hydrocarbons), nonvolatile organic compounds (NVOCs), and inorganic compounds and elements. Many of the 188 HAPs are not among the chemicals routinely measured in ambient air sampling programs for ozone precursors or toxic air pollutants. For example, only 70 of the 188 HAPs were included in the EPA's National Volatile Organic Compound Database,[1,2] a compilation of data on more than 300 commonly measured air pollutants prepared shortly before the 1990 CAA Amendments were promulgated. As a result, data with which to evaluate the potential public health risks from the 188 HAPs may not be readily available.

This chapter summarizes the results of a survey of ambient air concentrations of the 188 HAPs. To the extent possible, the definition of an ambient measurement used in this survey was that stated in section 3.1, i.e., a measurement in the open atmosphere away from direct source impacts, and suitable for assessing the pollutant exposure of the general population. Thus, the results of this survey should be useful in estimating public exposure to the HAPs. Just as important, the survey has identified significant gaps in our knowledge of the ambient levels of several HAPs. Filling these gaps should be given a high priority, so that the public health risks from these chemicals can be evaluated with a satisfactory degree of certainty.

4.2 SURVEY PROCEDURES

For the purposes of this survey, the 188 diverse chemicals designated as HAPs were organized according to the chemical classes and volatility classes identified in Chapter 2. This classification was useful because, as shown in Chapter 3, similar chemicals are frequently measured together, using similar measurement methods. The survey was conducted in two stages. In the first, information on ambient concentrations of the 188 HAPs was located through keyword searches of appropriate computerized databases, in review articles, reference books, proceedings of relevant air-quality conferences, and in unpublished datasets from urban air monitoring studies. The results of that first stage survey have been reported.[4,5]

Ambient concentrations for 70 of the 188 HAPs were compiled through 1987 in the National VOC Data Base[1,2], which was updated concurrently with the first stage of this survey.[3] For this survey, the ambient data in the 1988 version of the national database[1,2] were summarized and supplemented with ambient data from other measurement programs. The search strategy for the 118 HAPs not included in the National VOC Data Base differed somewhat from the 70 included. Those 118 chemicals were the subject of computerized and manual searches of the literature to locate ambient data. For each chemical, a keyword search was conducted through the computerized databases of STN International (Columbus, OH). The databases searched included the Chemical Abstracts (CA) files from 1967 to 1993, Chemical Abstracts Previews (CAP) current files, and National Technical Information Service (NTIS) files from 1964 to 1992. To focus on data pertinent to toxics exposure of the U.S. population, the search was restricted to English-language citations authored in the United States. The strategy used both abstract and basic index searches to increase the likelihood of finding relevant citations.

Master sets of literature citations were set up in each of the STN files searched. These master sets were then combined with the chemical names and CAS registry numbers of the compounds, to produce citation listings specific to each HAP. If the initial reviews indicated information of value, the listed citations were then reviewed by title, abstract, and in their entirety. For all of the 188 HAPs, data were obtained from published reviews, reference texts, and from proceedings of meetings such as the annual EPA/Air and Waste Management Association (AWMA) annual symposium on toxic air pollutants. By contacting the respective lead scientists, recent data were also obtained from unpublished field studies.

The list of 188 HAPs includes some redundant entries in the form of chemical groups (e.g., xylenes, cresols) and their individual constituent isomers. These chemicals may be used in industrial settings as the mixed isomers, but are generally measured in the atmosphere as individual isomers. Searches were performed for both the individual and mixed isomers, but ambient data were found primarily for the individual isomers. The HAP denoted as polycyclic organic matter (POM) comprises numerous individual compounds, and the compounds measured are not always clearly defined in reports of ambient measurements. For consistency, and to emphasize potential health risks from POM, this survey focused on eight individual POM compounds identified as possible or probable human carcinogens.[6,7] Those eight compounds are benzo[a]pyrene, benzo[a]anthracene, dibenzo[a,h]anthracene, chrysene, benzo[b]fluoranthene, benzo[k]fluoranthene, indeno[1,2,3,c-d]pyrene, and benzo[g,h,i]perylene. Ambient data were compiled for the sum of these eight POM compounds.

The second stage of this survey of ambient HAP concentrations relied upon a recently assembled database of ambient monitoring data.[8] Developed for the EPA's Office of Air Quality Planning and Standards (OAQPS), that database encompasses a larger number of chemical species, studies, and measurements than does the National VOC Database.[1-3] Although there is no federal mandate to do so, numerous state and local agencies conduct sampling programs for toxic air pollutants, including the 188 HAPs. To identify these sources of ambient data, EPA directed a search through a number of different means that included professional organizations such as the State and Territorial Air Pollutant Program Administrators (STAPPA), the Association of Local Air Pollution Control Officers (ALAPCO), and the AWMA, as well as Internet information provided on state environmental agencies and other contact referrals. Once these agencies were identified, cooperative sources contributed suitable ambient monitoring data to OAQPS. A database of these ambient concentration measurements was then compiled.[8]

To produce the most complete and comprehensive archive of ambient measurements, the data obtained from state and local monitoring efforts were combined with similar data from the Aerometric Information Retrieval System (AIRS). Administered by the OAQPS, AIRS is a computer-based repository of U.S. air pollution data. Data contributed to AIRS are largely the result of regulatory monitoring of criteria pollutants by state and local agencies. However, depending on the attainment status of a region, some non-criteria pollutant monitoring is mandatory, as is submission of the data to AIRS. All collected data are merged into the archive on a regular basis; the latest update occurred in the fall of 2000.

The ambient data archive[8] was made available for this survey by staff of Battelle's Statistics and Data Analysis Systems department, who are assembling a web-based, readily accessible version of the database. The database contains several sub-databases that link to one another by one or more common fields, eliminating redundant data and making the database smaller and easier to work with. Those sub-databases include information on the sampling program, sampling site, measurement method, pollutant identification, ambient pollutant concentration, and detection limit of the measurement method. Ambient concentration data were obtained from this database for 97 of the 188 HAPs.

The intent of this ambient concentration survey was not to catalog every data point or sample. Rather, the aim was to compile information on typical concentrations (i.e., mean and/or median), the range of concentrations observed, and the number, locations, and time periods of the measurements. The purpose of this approach was to provide concentration data suitable for estimating population exposures to the 188 HAPs. In general, the scarcer the ambient data for a given HAP, the greater the effort spent to find such data. Additional information such as the detection limit of the measurements and the number of results below the detection limit was also recorded when available.

In keeping with the aim of providing data for health risk assessment, the focus of this survey was on ambient data in populated (urban to rural) areas of the U.S. To that end, effort was made to exclude data from remote sites, and data indicating strong, direct, local source contributions. In some cases, such exclusion was called for by clear identification of the origin of the samples. However, in many cases, identification was ambiguous and, in the absence of clear information, elevated concentration results were generally retained in the dataset. There was no attempt to exclude measurements that may have been subject to some impact of local urban sources, because those data properly represent the upper range of concentrations to which urban residents may be exposed.

4.3 AMBIENT AIR CONCENTRATIONS OF HAPs

Ambient air concentrations of hazardous air pollutants are compiled in Table 4.1 (see Appendix following Chapter 4), which lists all 188 HAPs in the same order as in Title III of the CAA, with alternate names as well, if they were stated in the HAPs list. Table 4.1 gives the name and CAS number for each compound; the locations and years of measurements; the number of ambient measurements (N); the mean, range, and median (if available) of the measured data; the number of the pertinent reference in the associated reference list that follows Table 4.1; and additional comments on the data, such as the number of non-detects included in the reported data. The concentration units for each HAP are indicated in the first line of the concentration data. All concentrations are in mass per volume units, i.e., micrograms per cubic meter ($\mu g/m^3$), nanograms per cubic meter (ng/m^3), or picograms per cubic meter (pg/m^3). As noted in Chapter 3, mass per volume concentration units can be readily converted into mixing ratios at assumed atmospheric conditions. For example, at 20° C and one atmosphere pressure, the conversion between $\mu g/m^3$ concentrations and part-per-billion by volume (ppbv) mixing ratios is

$$1 \ \mu g/m^3 = \frac{1}{0.0416 \cdot MW} \ \text{ppb}$$

or

$$1 \ \text{ppbv} = 0.0416 \cdot MW \ \mu g/m^3$$

where MW is the molecular weight of the HAP in question.

The Year column in Table 4.1 indicates the period of data collection for each data source. Note that, in some cases, the number of locations and number of samples were not evident from the literature. In those cases, the numbers were estimated, or lower limits are shown. Inconsistency was also found in the treatment of measurements below the detection limit. Some studies failed to state the detection limit, or to define the number of measurements below that limit. The value assigned to non-detects (e.g., zero, half the detection limit, etc.) in calculating a mean value was also not always clearly stated. Whenever possible, these inconsistencies were addressed by inferring or estimating the detection limits and number of non-detects from information in the literature. Mean values were calculated using a value of one half the estimated detection limit for the results that were listed as non-detects.

The most noticeable feature of the data in Table 4.1 is the extremely wide variation in the amount of data found for individual HAPs. The number of sampling locations for individual HAPs varies from zero to more than 900 sites, and the number of measurements varies from zero to more than 470,000. Of particular importance is that the number of samples is zero for 60 of the HAPs, i.e., no ambient concentration data were found. These features of the HAPs data are presented in more detail in Figures 4.1 and 4.2 for the 186 HAPs for which the number of sampling locations and number of measurements could be established. (For fine mineral fibers and radionuclides, ambient concentration estimates were drawn from sources that did not allow specification of the numbers of sampling sites and measurements.) Figure 4.1 shows the frequency distribution of the HAPs by number of sampling locations. The greatest frequency is found for zero sampling locations, with the 60 HAPs in this category composing nearly one third of the HAPs list. The second-largest frequency in Figure 4.1 is for 1–10 sampling locations, again indicating the scarcity of data for some HAPs. Only 86 chemicals (46% of the list) show data from more than 10 locations, and only 49 (26%) show data from 50 or more locations.

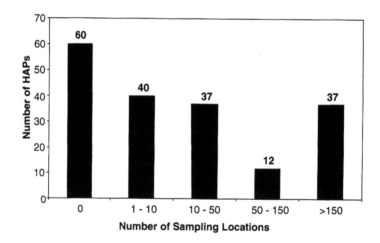

FIGURE 4.1 Distribution of the HAPs by number of ambient air sampling locations.

Figure 4.2 shows the corresponding frequency distribution by number of measurements found, and clearly indicates the wide range in the availability of ambient data for the HAPs. The 60 chemicals for which no ambient data were found constitute the largest frequency range in Figure 4.2. For a total of 83 chemicals (44% of the list), fewer than 100 measurements of each exist, and a total of 106 chemicals (56% of the HAPs list) show fewer than 1,000 measurements each. However, the second-largest frequency range includes the 36 chemicals for which between 10,000 and 100,000 measurements were found, and for nine HAPs, more than 100,000 measurements were found. These observations illustrate the primary characteristic of the HAPs list from the CAAA: it is a unique mix of some chemicals frequently measured in ambient air, and others rarely or never measured.

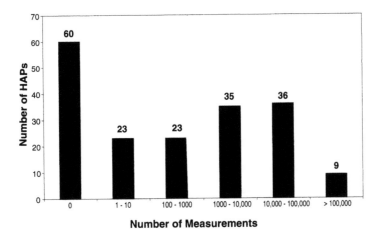

FIGURE 4.2 Distribution of the HAPs by number of ambient concentration measurements.

4.4 DATA GAPS

It is instructive to explore what types of chemicals predominate among those HAPs for which no ambient data were found. That subject is addressed in Figure 4.3, which shows the total number of HAPs and the number with no available data for each of the chemical categories established in Chapter 2 (Table 2.2). As expected, for categories such as hydrocarbons, aromatic compounds, and their halogenated analogs, data are available for all or nearly all of the HAPs. These HAPs are common toxic, relatively nonpolar VOCs, and are readily measured in ambient air by methods such as EPA Compendium Method TO-14.[9] In contrast, no data are available for most of the HAPs in the nitrogenated organic category, and for one third of the HAPs in the oxygenated organic category. This fact is particularly important because together, these two groups make up nearly half of the HAPs list (88 total HAPs).

Several reasons may exist for the scarcity of ambient measurements within some chemical categories. For the nitrogenated and oxygenated organics, which collectively fall under the definition of polar VOCs, the most likely reason is the lack of sampling and analysis methods for these compounds. Due to their water solubility and reactivity, measurement of these chemicals at likely ambient levels of a few $\mu g/m^3$ or less (ppbv to sub-ppbv mixing ratios) is more difficult than measurement of VOCs, and methods for many of these chemicals are still in development (see Chapter 3). That this development is occurring is confirmed by the substantially improved state of ambient data shown in Figure 4.3 for the nitrogenated and oxygenated organics, relative to the state at the time of the initial ambient concentration survey.[4,5] For example, Figure 4.3 shows that, of the 49 nitrogenated organics, 34 have no ambient concentration data, whereas, in the initial survey,[4,5] 39 of the nitrogenated organics had no ambient data. The corresponding results for the 39 oxygenated organics (caprolactam was dropped from the HAPs list since the initial survey) are 13 compounds with no data at this time, compared with 21 with no data in the initial survey.[4,5] For comparison, no change occurred in the number of HAPs with no ambient data in the chemical categories of hydrocarbons, halogenated hydrocarbons, halogenated aromatics, inorganics, pesti-

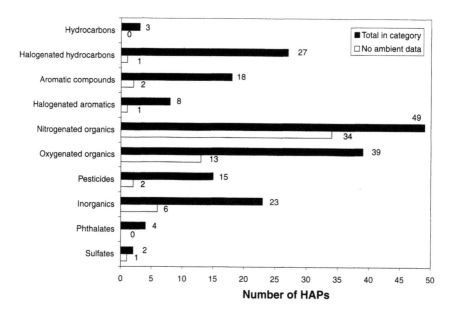

FIGURE 4.3 Number of HAPs, including the number with no ambient data, for each chemical category.

cides, phthalates, and sulfates. The category of aromatics (18 total HAPs) showed an improvement from four HAPs with no ambient data[4,5] to two with no ambient data at this time. It can be concluded that development and validation of measurement methods for polar volatile compounds in air is particularly needed before risk assessment and regulation of many of these HAPs can be adequately accomplished.

For other chemical categories, the scarcity of data may have other causes. Pesticide measurements, for example, are generally made in agricultural areas in association with application of these chemicals. Such measurements are not directly relevant to the exposure of the urban U.S. population and are not included in the tabulated data. Many other chemicals have been measured in the workplace but not in ambient air. For example, the list designates titanium tetrachloride, elemental phosphorus, and dye intermediates such as 3,3'-dimethoxybenzidine, as HAPs. Although the potential toxicity of these chemicals has been established, their ambient concentrations have not been measured because they have been considered unlikely to be present at significant concentrations except near very localized sources. For such compounds, assessment of the potential for human exposure might best be focused in areas around known sources.

Another reason for the lack of ambient air data for some HAPs is the ambiguous nature of the identification on the CAA list. A good example is coke oven emissions. The emission of a variety of toxic chemicals, including sulfur compounds, benzene, other aromatics, and polycyclic aromatic compounds, from coke ovens is well documented. However, it is impossible to quantify those compounds originating in ambient air from coke oven emissions in the face of other sources of the same compounds, without, for example, detailed source apportionment modeling in the area of a coke oven source. As a result, measurements of coke oven emissions as a chemical group in urban areas simply do not exist.

The representativeness of the HAPs data for use in health risk assessments is an important issue. Clearly, both the number of measurements and the number of locations in which measurements have been made are important in this regard (Table 4.1). Some HAPs, such as many of the chlorinated and aromatic hydrocarbons, have been measured tens of thousands of times in hundreds of locations. The geographic spread of these data is also wide, because of the large number of studies that have included these chemicals. Thus, it can be argued that sufficient data exist to estimate typical and elevated human exposures to these chemicals. However, as noted above, nearly half of the 188 HAPs have been measured fewer than 100 times, and more than half have been measured in fewer than 10 locations. Such small datasets and limited geographic coverage are unlikely to represent adequately the exposure of the U.S. population to those chemicals. For many of the HAPs, therefore, the representativeness of the existing data is very limited. More measurements of these compounds are needed for adequate health risk assessment.

TABLE 4.2
Summary of Selected Data for the 33 HAPs Designated as High Priority Urban Air Toxics

High Priority HAP	No. of Locations	No. of Measurements	Mean	Range	Years
Acetaldehyde	216	14,143	2.45 µg/m^3	< 0.004–102	1980–2000
Acrolein	66	4,379	0.27 µg/m^3	< 0.007–49.8	1980–2000
Acrylonitrile	22	2,926	0.27 µg/m^3	< 0.13–39	1990–2000
Arsenic cmpds.	32	8,431	1 ng/m^3	< 2–50	1990–1998
Benzene	783	350,221	2.15 µg/m^3	< 0.005–2836	1980–2000
Beryllium cmpds.	30	5,218	0.3 ng/m^3	< 0.001–18	1990–2000
1,3-butadiene	341	102,638	0.79 µg/m^3	< 0.022–314	1973–2000
Cadmium cmpds.	22	2,093	3 ng/m^3	2.5–17	1990–1992
Carbon tetrachloride	383	39,404	1.24 µg/m^3	< 0.006–1,493	1975–2000
Chloroform	496	57,258	0.81 µg/m^3	< 0.002–4,334	1981–2000
Chromium cmpds.	118	70,917	1 ng/m^3	< 0.4–95	1988–1999
Coke oven emissions*	--	--	--	--	--
1,3-dichloropropene	15	392	0.18 µg/m^3	ND (< 0.45)	1993–1994
2,3,7,8-TCDD	10	72	0.009 pg/m^3	< 0.001–0.022	1987–1995
Ethylene dibromide	271	27,269	0.89 µg/m^3	< 0.038–962	1983–1999
Ethylene dichloride	427	49,908	1.34 µg/m^3	< 0.008–2,873	1981–2000
Ethylene oxide	2	< 10 (estimated)	0.09 µg/m^3	0.05–1.8	1989
Formaldehyde	250	51,801	6.39 µg/m^3	< 0.003-436	1977–2000
Hexachlorobenzene	10	1,023	2 ng/m^3	ND (< 6)	1992–1997
Hydrazine	--	--	--	--	--
Lead cmpds.	117	69,606	2 ng/m^3	< 0.06 -320	1988–1999
Manganese cmpds.	118	70,796	1 ng/m^3	< 0.4–113	1988–1999
Mercury cmpds.	3	15,546 (vapor)	1.86 µg/m^3	0.5-10.1	1988-1998
	3	391 (particle)	0.082 µg/m^3	0.005-0.75	
Methylene chloride	481	54,405	2.4 µg/m^3	< 0.035-2,812	1983–2000
Nickel cmpds.	118	70,798	0.3 ng/m^3	< 0.05–55	1988–1999
Polychlorinated biphenyls	10	--	5 ng/m^3	0.5–36	1973–1978
Polycyclic Organic Matter*	11	129	8.7 ng/m^3	0.3–91	1988–1991
Propylene dichloride	336	30,842	0.50 µg/m^3	< 0.009-118	1983–2000
Quinoline	2	3	0.34 µg/m^3	up to 1.0	1982
1,1,2,2-Tetrachloroethane	232	13,457	0.27 µg/m^3	< 0.014-38.6	1983-1999
Tetrachloroethylene	482	46,899	1.65 µg/m^3	< 0.034-6,020	1981–2000
Trichloroethylene	426	48,411	0.96 µg/m^3	< 0.011–321	1981–2000
Vinyl chloride	365	40,094	0.41 µg/m^3	< 0.026–125	1986–2000

*See text for discussion of the composition of these chemical groups.

4.5 RECENT DATA FOR HIGH PRIORITY HAPs

As a final example of the ambient HAPs concentration data, Table 4.2 summarizes selected ambient data for the group of 33 high priority HAPs identified in Chapter 1. The data shown in Table 4.2 are a subset of the complete datasets compiled for these chemicals and shown in Table 4.1. Shown in Table 4.2 are the number of study locations, number of samples, mean, range, and years of recent measurements for the 33 high priority HAPs.

The availability of data for the 33 HAPs in Table 4.2 is generally better than for the 188 HAPs as a whole. For most of these HAPs, substantial numbers of recent samples are indicated. Exceptions are coke oven emissions and hydrazine, for which no ambient data exist, and ethylene oxide and quinoline, for which only a few ambient measurements were found. Data are also relatively scarce for 1,3-dichloropropene, 2,3,7,8-TCDD, hexachlorobenzene, particulate mercury compounds, and polycyclic organic matter. Recent ambient measurements of polychlorinated biphenyls are also scarce. With these exceptions, Table 4.2 shows that ambient data exist with which to estimate population exposures for the majority of the 33 high priority HAPs. Inspection of the full dataset (Table 4.1) also suggests that the recent data in Table 4.2 exhibit means and ranges that are generally lower than those of earlier data. This difference may indicate decreases in the emissions of these chemicals. However, changes in the choice of sampling locations might also account for this difference. Site selection in early urban field studies often emphasized worst-case locations such as urban traffic centers, whereas recent studies have tended to emphasize sites that are more representative of local population distributions. As a result, the recent data shown in Table 4.2 may be useful for initial human exposure assessments for these 33 HAPs.

4.6 SUMMARY

This chapter has presented an updated assessment of the ambient concentration data available for the 188 HAPs. The primary observation to be made is that the HAPs list includes a large group of chemicals that have rarely or never been measured in ambient air, and another group that has been measured very frequently. For example, 60 of the 188 HAPs (nearly one third of the list) have no ambient concentration data, and 83 HAPs (44% of the list) have been measured fewer than 100 times. On the other hand, 45 HAPs (24% of the listed HAPs) have been determined more than 10,000 times each in ambient air. These results indicate that the representativeness of existing ambient data for estimating population exposures to HAPs will vary widely among different HAPs.

A somewhat more optimistic picture emerges when considering the 33 HAPs considered to cause the greatest public health risks in urban areas. Substantial numbers of ambient data are available for most of the 33 high priority HAPs, and thus, for most of these 33 HAPs, estimates of population exposures should be feasible.

REFERENCES

1. Shah, J.J. and Heyerdahl, E.K., National Ambient Volatile Organic Compounds (VOCs) Data Base Update, Report EPA-600/3-88/010(a), U.S. EPA, Research Triangle Park, NC, 1988.
2. Shah, J.J. and Singh, H.B., Distribution of volatile organic chemicals in outdoor and indoor air: A national VOCs database, *Environ. Sci. Technol.*, 22, 1381, 1988.
3. Shah, J.J. and Joseph, D.W., National VOC Data Base Update 3.0, Final Report to U.S. EPA, EPA-600/R-94-089, G2 Environmental, Inc., Washington, D.C., May 1993.
4. Kelly, T.J. et al., Ambient Concentration Summaries for Clean Air Act Title III Hazardous Air Pollutants, Final Report to U.S. EPA, EPA-600/R-94-090, Battelle, Columbus, OH, July 1993.
5. Mukund, R. et al, Status of ambient measurement methods for hazardous air pollutants, *Environ. Sci. Tech.*, 29, 183, 1995.

6. Menzie, C.A., Potocki, B.B, and Santodonato, J., Exposure to carcinogenic PAHs in the environment, *Environ. Sci. Technol.*, 26, 1278, 1992.
7. IARC Monographs on the Evaluation of the Carcinogenic Risk of Chemicals to Humans: Polynuclear Aromatic Hydrocarbons, Part 1, Chemical, Environmental, and Experimental Agency for Research on Cancer, World Health Organization, 1983.
8. Rosenbaum, A.S., Stiefer, P.S. and Iwamiya, R.K., Air Toxics Data Archive and AIRS Combined Data Set: Data Base Descriptions, prepared for U.S. EPA, Office of Policy, Planning, and Evaluation, by Systems Applications International, SYSAPP-99/25, 1999.
9. McClenny, W.A. et al., Canister-based method for monitoring toxic VOCs in ambient air, *J. Air Waste Mgt. Assoc.*, 41, 1308, 1991.

APPENDIX

TABLE 4.1
Ambient Air Concentrations of 188 Hazardous Air Pollutants (Chemicals shown in italics are high priority urban HAPs)

Compound and CAS Number	Locations	Year	N[a]	Concentration[b] Mean	Concentration[b] Range	Ref	Comments
Acetaldehyde	*216 U.S. locations*	*1980-00*	*14,143*	*2.446 µg/m³*	*< 0.004-102 µg/m³*	*28*	*Includes 735 nondetects*
75-07-0	*34 U.S. urban to suburban locations*	*1974-84*	*384*	*6.09*	*up to 105*	*71*	*Includes 189 nondetects*
	Lima, OH	*1990-91*	*56*	*3.2*	*< 0.2-16.7*	*21*	*Includes 19 nondetects*
	Columbus, OH	*1989*	*332*	*2.32*	*0.37-16.7*	*33*	
	14 U.S. urban sites	*1989*	*406*	*2.51*	*0.68-13.9*	*78*	
				(Median 2.21)	*(Site means 1.72-3.40)*		
Acetamide 60-35-5	—	—	—	—	—	—	—
Acetonitrile	2 U.S. locations	1990-91	44	0.84 µg/m³	< 1.68 µg/m³	20	All nondetects
75-05-8	2 U.S. urban to suburban locations	1982	4	0.05	up to 0.16	71	Includes 2 nondetects
	Lima, OH	1990-91	8	0.84	< 1.68	21	All nondetects
Acetophenone	3 U.S. locations	1993	3	17.5 µg/m³	16.2-18.7 µg/m³	28	
98-86-2	2 U.S. urban locations	1977-78	3	0.15	up to 0.30	71	Includes 1 nondetect
2-Acetylaminofluorene 53-96-3	—	—	—	—	—	—	—
Acrolein	*66 U.S. locations*	*1980-00*	*4,379*	*0.272 µg/m³*	*< 0.007-49.8 µg/m³*	*28*	*Includes 3,882 nondetects*
107-02-8							

Compound / CAS No.	Location	Period	N	Mean (Median)	Range	Ref.	Notes
Acrylamide 79-06-1	—	—	—	—	—	—	—
Acrylic Acid 79-10-7	—	—	—	—	—	—	—
Acrylonitrile 107-13-1	22 U.S. locations	1990-00	2,926	0.273 µg/m³	< 0.130-39 µg/m³	28	Includes 2,714 nondetects
	4 U.S. urban to suburban locations	1981	36	0.66 (Median 0.22)	up to 4.42	71	Includes 4 nondetects
	Houston, TX	1990-91	22	1.1	< 2.2	20	All nondetects
	Boston, MA	1990-91	22	1.1	< 2.2	20	All nondetects
	Lima, OH	1990-91	8	1.1	< 2.2	21	All nondetects
Allyl chloride (3-chloro-1-propene) 107-05-1	32 U.S. locations	1988-98	1,261	0.266 µg/m³	< 0.156-2.57 µg/m³	28	Includes 1,146 nondetects
	Lima, OH	1990-91	81	0.16	< 0.32	21	All nondetects
	5 U.S. cities	1980-81	?	Range of site means < 0.016-0.019	< 0.016-0.060	77	
4-Aminobiphenyl 92-67-1	—	—	—	—	—	—	
Aniline 62-53-3	10 U.S. locations	1992-97	1,023	0.003 µg/m³	< 0.008 µg/m³	28	All nondetects
	3 U.S. urban to suburban locations	1979-82	4	1.3	0-5.1	71	Includes 3 nondetects
o-Anisidine 90-04-0	—	—	—	—	—	—	
Asbestos 1332-21-4	U.S. urban areas	1970s	-	-	1-100 ng/m³ (roughly 30-3000 fibers/m³) 1-70	54	
	U.S. urban areas	1970s	-	(Median 20)	Up to 9.8 (90th %) (roughly up to 330 fibers/m³)	55	
	U.S. urban areas	ca.1980	31	(Median 0.9)	Up to 6.8 (90th %)	56	
	48 U.S. cities	ca.1970	187	(Median 1.6)	(roughly up to 230 fibers/m³)	57	

TABLE 4.1 (CONTINUED)
Ambient Air Concentrations of 188 Hazardous Air Pollutants (Chemicals shown in italics are high priority urban HAPs)

Compound and CAS Number	Locations	Year	N^a	Concentration[b]		Ref	Comments
				Mean	Range		
Benzene	*783 U.S. locations*	*1980-00*	*350,221*	*2.147 µg/m³*	*< 0.005-2836 µg/m³*	*28*	*Includes 23,054 nondetects*
71-43-2	*127 U.S. urban to suburban locations*	*1973-87*	*5,756*	*10.1 (Median 6.5)*	*0-206*	*71*	*Includes 136 nondetects*
	Columbus, OH	*1989*	*298*	*1.56*	*0.10-5.30*	*33*	
	Atlanta, GA	*1990*	*4,620*	*2.68 (Median 1.92)*	*< 0.05-21.7*	*73*	
	11 U.S. cities	*1990*	*349*	*4.77 (Median 3.3)*	*0.13-67.3*	*74*	
	Lima, OH	*1990-91*	*81*	*2.56*	*1.10-6.75*	*21*	
Benzidine 92-87-5	10 U.S. locations	1992-97	1,023	0.015 µg/m³	< 0.006-0.0235 µg/m³	28	Includes 1,012 nondetects
Benzotrichloride 98-07-7	—	—	—	—	—	—	—
Benzyl chloride 100-44-7	114 U.S. locations	1988-99	8,174	0.249 µg/m³	< 0.052-8.5 µg/m³	28	Includes 7,716 nondetects
	4 U.S. urban to suburban locations	1980	42	0.05 (Median 0.05)	up to 0.32	71	Includes 2 nondetects
	Columbus, OH	1989	298	1.84	< 0.16-8.42	33	Includes 119 nondetects
	Lima, OH	1990-91	81	0.27	< 0.53	21	All nondetects
Biphenyl 92-52-4	16 U.S. locations	1993-97	811	32 ng/m³	2-2,138 ng/m³	28	Includes 548 nondetects
	Columbia, SC	1989	2	14.8	13.9-15.7	17	

Compound / CAS	Location	Year					Comments
Bis (2-ethylhexyl) phthalate (DEHP) 117-81-7	11 U.S. locations	1992-97	984	5 ng/m³	< 1-310 ng/m³	28	Includes 836 nondetects
	New York City, NY	1978	?	—	10.2-16.8	15	3 urban sites
	College Station, TX	1979-80	14	1.99	0.77-3.60	1	Rural site
	Gulf Coast, TX	1982	?	0.62	—	9	
	Portland, OR	1984-85	10	0.39 (gas-phase) 0.48 (particle-phase)	—	32	Number of samples estimated
Bis (chloromethyl) ether 542-88-1	—	—	—	—	—	—	—
Bromoform 75-25-2	3 U.S. urban to suburban locations	1976-79	26	60 ng/m³	up to 75 ng/m³	71	Includes 17 nondetects
	14 U.S. urban sites	1989	397	5	< 10-320	79	Includes 395 nondetects
1,3-Butadiene 106-99-0	*341 U.S. locations*	*1973-00*	*102,638*	*0.794 µg/m³*	*< 0.022-314 µg/m³*	*28*	*Includes 27,017 nondetects*
	46 U.S. urban to suburban locations	*1968-87*	*819*	*5.68*	*up to 332*	*71*	*Includes 209 nondetects*
	11 U.S. cities	*1990*	*349*	*2.3 (Median 0.53)*	*< 0.2-321*	*74*	*Includes 243 nondetects*
Calcium cyanamide 156-62-7	—	—	—	—	—	—	—
Captan 133-06-2	Jacksonville, FL	1987-88	47	0.8 ng/m³	< 1.6 ng/m³	11	Summer samples; All nondetects
			72	1.3	< 2.5	11	Spring samples; All nondetects
			70	3.1	< 6.1	11	Winter samples; All nondetects
	Springfield/Chicopee, MA	1987-88	49	2.7	< 5.4	11	Spring samples; All nondetects
			50	2.3	< 4.5	11	Winter samples; All nondetects Detection limits estimated

TABLE 4.1 (CONTINUED)
Ambient Air Concentrations of 188 Hazardous Air Pollutants (Chemicals shown in italics are high priority urban HAPs)

Compound and CAS Number	Locations	Year	Nª	Concentration[b] Mean	Range	Ref	Comments
Carbaryl (Sevin) 63-25-2	Jacksonville, FL	1987-88	47	4.6 ng/m³	< 9-11 ng/m³	11	Summer samples; Includes 46 nondetects
			72	12.5	< 25	11	Spring samples; All nondetects
			70	4	< 8	11	Winter samples; All nondetects
	Springfield/Chicopee, MA	1987-88	49	12.5	< 25	11	Spring samples; All nondetects
			50	5.5	< 11	11	Winter samples; All nondetects. Detection limits estimated
Carbon disulfide 75-15-0	42 U.S. locations	1991-99	436	10.7 µg/m³	< 0.003-654 µg/m³	28	Includes 242 nondetects
	3 U.S. urban to suburban locations	1979-82	15	0.3 (Median 0.14)	0.05-1.07	71	
Carbon tetrachloride 56-23-5	*383 U.S. locations*	*1975-00*	*39,404*	*1.24 µg/m³*	*< 0.006-1,493 µg/m³*	*28*	*Includes 20,811 nondetects*
	118 U.S. urban to suburban locations	*1972-87*	*5,011*	*0.96 (Median 0.77)*	*up to 38.8*	*71*	*Includes 1,014 nondetects*
	Columbus, OH	*1989*	*298*	*0.77*	*0.38-1.09*	*33*	*Includes 3 nondetects*
	11 U.S. cities	*1990*	*349*	*1.60*	*< 0.06-27.8*	*74*	*Includes 5 nondetects*
	Lima, OH	*1990-91*	*81*	*0.64 (Median 0.96)*	*< 0.64-0.83*	*21*	
Carbonyl sulfide 463-58-1	4 U.S. urban to suburban locations	1977-82	18	1.2 µg/m³ (Median 1.3)	1.0-1.4 µg/m³	71	
Catechol 120-80-9	8 U.S. locations	1994	29	4.35 µg/m³	< 9	28	All nondetects

Compound / CAS	Location	Years	No.	Concentration	Range		Comments
Chloramben 133-90-4	—	—	—	—	—	—	—
Chlordane 57-74-9	Miami, FL	1973-74	14	0.1 ng/m³	up to 0.9 ng/m³ (α-chlordane)	82	13 non-detected samples (each isomer); urban site
			14	0.1	up to 1.4 (γ-chlordane)		
	College Station, TX	1979-80	16	1.05	0.32-2.64	1	Rural sites
	Denver, CO	1980	?	0.063	—	8	Downtown, winter
	Columbia, SC	1977-79	?	1.30	—	8	
	North Inlet, SC	1977-79	?	0.15	—	12	
	Gulf Coast, TX	1982	?	0.036	—	9	
	Jacksonville, FL	1987-88	60	38.4	up to 628	11	Summer samples
			72	9.5	up to 266	11	Spring samples
			70	27.4	up to 175	11	Winter samples
	Springfield/Chicopee, MA	1987-88	49	3.1	up to 75	11	Spring samples
			50	2.0	up to 89	11	Winter samples
Chlorine 7782-50-5	—	—	—	—	—	—	—
Chloroacetic acid 79-11-8	—	—	—	—	—	—	—
2-Chloroacetophenone 532-27-4	—	—	—	—	—	—	—
Chlorobenzene 108-90-7	420 U.S. locations	1981-00	44,355	0.68 µg/m³	< 0.025-643 µg/m³	28	Includes 38,893 nondetects
	49 U.S. urban to suburban locations	1976-86	1,012	2.15 (Median 0.84)	up to 99	71	Includes 154 nondetects
	Columbus, OH	1989	298	0.07	< 0.14	33	All nondetects
	11 U.S. cities	1990	349	0.14	< 0.09-9.1	74	Includes 300 nondetects
	Lima, OH	1990-91	81	0.24	< 0.47	21	All nondetects
Chlorobenzilate 510-15-6	—	—	—	—	—	—	—

TABLE 4.1 (CONTINUED)
Ambient Air Concentrations of 188 Hazardous Air Pollutants (Chemicals shown in italics are high priority urban HAPs)

Compound and CAS Number	Locations	Year	N[a]	Mean	Range	Ref	Comments
Chloroform 67-66-3	*496 U.S. locations*	*1981-00*	*57,258*	*0.81 µg/m³*	*< 0.002-4,334 µg/m³*	*28*	*Includes 41,268 nondetects*
	104 U.S. urban to suburban locations	*1973-87*	*3,640*	*2.68 (Median 0.27)*	*up to 145*	*71*	*Includes 859 nondetects*
	Columbus OH	*1989*	*298*	*0.16*	*< 0.15-0.89*	*33*	*Includes 250 nondetects*
	11 U.S. cities	*1990*	*349*	*0.55*	*< 0.03-115*	*74*	*Includes 332 nondetects*
	Lima, OH	*1990-91*	*81*	*0.25*	*< 0.50*	*21*	*All nondetects*
Chloromethyl methyl ether 107-30-2	—	—	—	—	—	—	—
Chloroprene (2-chloro 1,3-butadiene) 126-99-8	154 U.S. locations	1987-00	5,645	0.44 µg/m³	< 0.109-32.3 µg/m³	28	Includes 4,883 nondetects
	1 U.S. urban location	1976	1	2.2	—	71	
	11 U.S. cities	1990	349	0.29	< 0.19-5.66	74	Includes 261 nondetects
Cresol/cresyllic acid 1319-77-3	—	—	—	—	—	—	—
o-Cresol 95-48-7	10 U.S. locations	1992-97	1,015	0.005 µg/m³	< 0.001-0.532 µg/m³	28	Includes 977 nondetects
	Portland, OR	1984	7	0.07	up to 0.13	3	
m-Cresol 108-39-4	1 U.S. location	1993	1	11.5 µg/m³	—	28	
	2 U.S. urban locations	1982	3	1.4	up to 4.1	71	Includes 2 nondetects
p-Cresol 106-44-5	10 U.S. locations	1992-97	1,012	0.014 µg/m³	< 0.001-1.721 µg/m³	28	Includes 964 nondetects
	11 urban sites, CA	1985	62	4.6	0.5-19.9	19	
				2.1-8.6 (range of site means)			

	Year	Samples	Mean	Range	Ref	Comments
Cumene (isopropyl benzene) 98-82-8						
268 U.S. locations	1973-00	409,116	0.139 µg/m³	< 0.005-99 µg/m³	28	Includes 262,689 nondetects
48 U.S. urban to suburban locations	1976-86	938	1.3 (Median 0.25)	up to 141	71	Includes 305 nondetects
Atlanta, GA	1990	4620	0.34 (Median 0.18)	< 0.06-8.3	73	
2,4-D, salts and esters						
Baltimore, MD, Fresno, CA, Riverside, CA, SL City, UT	1967-68	437	–	up to 4.0 ng/m³	7,82	Includes 436 nondetects; detected in 1 sample at SL City, UT; detection limit unknown; urban sites
Jordan, NY	1980	?	1.15 ng/m³	–	13	One-yr study, 16 U.S. cities, 3 samples contained 2,4-D
Rome, NY	1980	?	1.54	–	13	
Salt Lake City, UT	1980	?	4.0	–	13	
Jacksonville, FL	1987-88	47	0.27	< 0.5-1.2	11	Summer samples; Includes 46 nondetects
		72	7.5	< 15	11	Spring samples; all nondetects
		70	5.5	< 11	11	Winter samples; all nondetects
Springfield/Chicopee, MA	1987-88	49	12	< 24	11	Spring samples; all nondetects
		50	7	< 14	11	Winter samples; all nondetects
						Detection limits estimated.

TABLE 4.1 (CONTINUED)
Ambient Air Concentrations of 188 Hazardous Air Pollutants (Chemicals shown in italics are high priority urban HAPs)

Compound and CAS Number	Locations	Year	N[a]	Concentration[b]		Ref	Comments
				Mean	Range		
DDE, 72-55-9	10 U.S. locations	1992-97	1,023	2 ng/m³	< 6 ng/m³	28	All nondetects
	Baltimore, MD, Fresno, CA, Riverside, CA, SL City, UT	1967-68	437	—	ND-6.4 (4,4'-DDE) ND (2,4'-DDE)	7,82	Includes 424 nondetects (4,4'); detection limits unknown; urban sites
	Buffalo, NY, Dothan, AL, Iowa City, IA, Orlando, FL, Stoneville, MS	1967-68	438	—	ND-131 (4,4'-DDE) ND-9.6 (2,4'-DDE)	7,82	Includes 291 undetects (4,4'; Includes 393 nondetects (2,4'); detection limits unknown, rural sites
	Columbia, SC	1978	?	4.5	up to 14.2	8	
	Denver, CO	1980	?	0.093	—	8	
	College Station, TX	1979-80	16	0.021	—	1	Rural site
	Jacksonville, FL	1987-88	72	0.26	0.04-0.66	11	Spring samples
		1987-88	70	—	—	11	Winter samples; all nondetects; detection limit unknown
	Springfield/Chicopee, MA	1987-88	49	—	—	11	Spring samples; all nondetects; detection limit unknown
			50	—	—	11	Winter samples; all nondetects; detection limit unknown
Diazomethane 334-88-3	—	—	—	—	—	—	—
Dibenzofuran 132-64-9	10 U.S. locations	1993-97	695	8 ng/m³	< 2-58 ng/m³	28	Includes 381 nondetects

Compound / CAS	Location	Year	n	Mean	Range	Ref	Notes
1,2-Dibromo-3-chloropropane 96-12-8	1 U.S. urban location	1976	3	0.01 µg/m³	< 0.01-0.02 µg/m³	71	Includes 2 nondetects
Dibutyl phthalate 84-74-2	11 U.S. locations	1992-97	1,012	3 ng/m³	< 2-101 ng/m³	28	Includes 912 nondetects
	New York City, NY	1978	?	–	3.3-5.7	15	3 urban sites
	College Station, TX	1979-80	13	1.40	0.48-3.60	1	Rural site
	Gulf Coast, TX	1982	?	0.42	–	9	
1,4-Dichlorobenzene (p-dichlorobenzene) 106-46-7	389 U.S. locations	1981-00	31,805	0.676 µg/m³	< 0.002-1,860 µg/m³	28	Includes 20,306 nondetects
	44 U.S. urban to suburban locations	1976-86	719	4.16 (Median 0.55)	up to 321	71	Includes 136 nondetects
	Columbus, OH	1989	298	0.09	< 0.18	33	All nondetects
	11 U.S. cities	1990	349	1.04	< 0.55-23.6	74	Includes 219 nondetects
	Lima, OH	1990-91	81	0.31	< 0.61	21	All nondetects
3,3'-Dichlorobenzidine 91-94-1	13 U.S. locations	1990-99	2,529	0.222 µg/m³	< 0.001-48 µg/m³	28	Includes 2,480 nondetects
Dichloroethyl ether (bis[2-chloroethyl] ether) 111-44-4	10 U.S. locations	1992-97	1,023	0.005 µg/m³	< 0.019 µg/m³	28	All nondetects
1,3-Dichloropropene 542-75-6	*15 U.S. locations NJ, D.C., FL, IL, OH, TX, LA, KS; Urban sites (Average over all sites)*	*1993-94*	*392*	*0.183 µg/m³*	*< 0.454 µg/m³*	*28*	*All nondetects trans- and cis- reported separately and summed; trans: includes 335 nondetects; cis: includes 339 nondetects*
	1990		*349*	*0.32*	*< 0.19-18.2*	*2*	
	Lima, OH	*1990-91*	*81*	*0.23*	*< 0.46*	*21*	*Sum of cis- and trans-isomers. All nondetects*

TABLE 4.1 (CONTINUED)
Ambient Air Concentrations of 188 Hazardous Air Pollutants (Chemicals shown in italics are high priority urban HAPs)

Compound and CAS Number	Locations	Year	N^a	Concentration[b]			Ref	Comments
				Mean	Range			
Dichlorvos 62-73-7	Jacksonville, FL	1987-88	47	0.75 ng/m³	< 1.5 ng/m³		11	Summer samples; all nondetects
			72	2.0	< 3.9		11	Spring samples; all nondetects
	Springfield/Chicopee, MA	1987-88	70	3.2	< 3.1-148		11	Winter samples
			49	2.8	< 5.6		11	Spring samples; all nondetects
			50	2.0	< 4.0		11	Winter samples; all nondetects
								Detection limits estimated
Diethanolamine 111-42-2	—	—	—	—	—		—	—
Diethyl sulfate 64-67-5	—	—	—	—	—		—	—
3,3'-Dimethoxybenzidine 119-90-4	—	—	—	—	—		—	—
4-Dimethylaminoazo- benzene 60-11-7	—	—	—	—	—		—	—
N,N-Dimethylaniline 121-69-7	—	—	—	—	—		—	—
3,3' Dimethylbenzidine 119-93-7	—	—	—	—	—		—	—
Dimethylcarbamoyl chloride 79-44-7	—	—	—	—	—		—	—

Chemical (CAS No.)	Location	Dates	No. of samples	Concentration		Concentration	Ref.	Comments
N,N-Dimethylformamide 68-12-2	Unspecified location, Northeast U.S. (suspected)	1983	4	9.8 µg/m³	—	< 0.02-13.8 µg/m³	23	Residential areas around waste site (unsettled wind)
			1	0.06		—	23	Upwind of waste site
1,1-Dimethylhydrazine 57-14-7	—	—	—	—	—	—	—	—
Dimethyl phthalate 131-11-3	10 U.S. locations	1992-97	1,000	0.006 µg/m³		< 0.001-0.113 µg/m³	28	Includes 954 nondetects
	Neenah, WI	1985-86	3	0.03		< 0.06	4	All nondetects
	Newark, NJ	1987	3	0.03		< 0.06	4	All nondetects
Dimethyl sulfate 77-78-1	1 U.S. urban location	1983	3	7.4 µg/m³ (Median 8.3)		2.5-11.6 µg/m³	71	
4,6-Dinitro-o-cresol, and salts N/A	10 U.S. locations	1992-97	1,023	0.007 µg/m³		< 0.021 µg/m³	28	All nondetects
2,4-Dinitrophenol 51-28-5	10 U.S. locations	1992-97	1,023	0.011 µg/m³		< 0.042 µg/m³	28	All nondetects
2,4-Dinitrotoluene 121-14-2	10 U.S. locations	1992-97	1,023	0.003 µg/m³		< 0.008 µg/m³	28	All nondetects
1,4-Dioxane 123-91-1	5 U.S. locations	1984-99	59	2.829 µg/m³		< 0.490-39.6 µg/m³	28	Includes 37 nondetects
	12 U.S. urban to suburban locations	1979-84	533	0.44		up to 30	71	Includes 346 nondetects
1,2-Diphenylhydrazine 122-66-7	—	—	—	—	—	—	—	—
Epichlorohydrin 106-89-8	8 U.S. locations	1994	32	2.110		< 5	28	All nondetects
1,2-Epoxybutane 106-88-7	—	—	—	—	—	—	—	—

TABLE 4.1 (CONTINUED)
Ambient Air Concentrations of 188 Hazardous Air Pollutants (Chemicals shown in italics are high priority urban HAPs)

Compound and CAS Number	Locations	Year	N[a]	Concentration[b]		Ref	Comments
				Mean	Range		
Ethyl acrylate 140-88-5	Lima, OH	1990-91	8	0.42 µg/m³	< 0.83 µg/m³	21	Two urban sites, all nondetects
Ethylbenzene 100-41-4	703 U.S. locations	1973-00	463,821	1.019 µg/m³	< 0.002-3874 µg/m³	28	Includes 96,608 nondetects
	79 U.S. urban to suburban locations	1968-87	3375	7.9 (Median 2.5)	up to 1,248	71	Includes 182 nondetects
	Columbus, OH	1989	298	1.10	0.13-5.56	33	
	Atlanta, GA	1990	4620	2.04 (Median 1.30)	< 0.06-36.7	73	
	11 U.S. cities	1990	349	1.68 (Median 1.06)	< 0.09-18.5	74	
	Lima, OH	1990-91	81	1.1	< 0.44-3.0	21	
Ethyl carbamate 51-79-6	—	—	—	—	—	—	—
Ethyl chloride (Chloroethane) 75-00-3	225 U.S. locations	1983-99	13,513	0.246 µg/m³	< 0.021-56.5 µg/m³	28	Includes 12,402 nondetects
	22 U.S. urban to suburban locations	1980-86	180	55.7 (Median 0.17)	up to 1,045	71	Includes 28 nondetects
	Columbus, OH	1989	298	0.04	< 0.08	33	All nondetects
	Lima, OH	1990-91	81	0.24	< 0.27-5.6	21	Includes 79 nondetects
	11 U.S. cities	1990	349	0.15	< 0.27-1.18	74	Includes 342 nondetects
Ethylene dibromide (1,2-Dibromoethane) 106-93-4	*271 U.S. locations*	*1983-99*	*27,269*	*0.893 µg/m³*	*< 0.038-962 µg/m³*	*28*	*Includes 25,903 nondetects*
	64 U.S. urban to suburban locations	*1976-87*	*2,120*	*1.80*	*up to 231*	*71*	*Includes 1,517 nondetects*
	Lima, OH	*1990-91*	*81*	*0.39*	*< 0.78*	*21*	*All nondetects*

Chemical / CAS	Location	Years	N	Mean	Range	%	Comments
Ethylene dichloride (1,2-dichloroethane) 107-06-2	427 U.S. locations	1981-00	49,908	1.340 µg/m³	< 0.008-2,873 µg/m³	28	Includes 43,496 nondetects
	70 U.S. urban to suburban locations	1976-87	2019	1.61	up to 74.7	71	Includes 883 nondetects
	Columbus OH	1989	298	0.16 (Median 0.04)	< 0.12-2.39	33	Includes 250 nondetects
	11 U.S. cities	1990	349	0.08	< 0.16-0.33	74	Includes 348 nondetects
	Lima, OH	1990-91	81	0.21	< 0.41	21	All nondetects
Ethylene glycol 107-21-1	6 U.S. locations	1994	14	78.6 µg/m³	< 200 µg/m³	28	All nondetects
Ethyleneimine 151-56-4	—	—	—	—	—	—	—
Ethylene oxide 75-21-8	San Luis Obispo, CA	1989	1	0.09 µg/m³	–	80	Urban site; Also reported: remote marine air: 0.027-0.046; and urban near-source: >1.8
Ethylene thiourea 96-45-7	Los Angeles, CA	1989	?	–	0.05-1.8	81	—
Ethylidene dichloride (1,1-dichloroethane) 75-34-3	326 U.S. locations	1987-00	30,460	0.228 µg/m³	< 0.040-50.2 µg/m³	28	Includes 29,757 nondetects
	31 U.S. urban to suburban locations	1976-87	644	0.17	up to 8.2	71	Includes 437 nondetects
	Columbus, OH	1989	298	0.06	< 0.12	33	All nondetects
	11 U.S. cities	1990	349	< 0.16	< 0.16-3.75	74	Includes 335 nondetects
	Lima, OH	1990-91	81	0.21	< 0.41	21	All nondetects
Formaldehyde 50-00-0	250 U.S. locations	1977-00	51,801	6.387 µg/m³	< 0.003-436 µg/m³	28	Includes 1,005 nondetects
	42 urban to suburban locations	1972-87	554	8.93 (Median 4.71)	up to 87.1	71	Includes 75 nondetects
	Columbus, Ohio	1989	332	3.74	0.12-19.2	33	
	14 U.S. urban sites	1989	416	2.64 (Median 2.22)	0.53-11.0 (Site means 1.75-4.72)	78	

TABLE 4.1 (CONTINUED)
Ambient Air Concentrations of 188 Hazardous Air Pollutant (Chemicals shown in italics are high priority urban HAPs)

Compound and CAS Number	Locations	Year	N[a]	Concentration[b] Mean	Concentration[b] Range	Ref	Comments
Heptachlor 76-44-8	10 U.S. locations	1992-97	1,023	0.002 µg/m³	< 0.006 µg/m³	28	All nondetects
	Miami, FL	1973-74	14	1.2	up to 2.2	82	Not detected in 1 sample; urban site
	Jacksonville, FL	1987-88	60	30.2	up to 627	11	Summer samples
			72	10.7	up to 220	11	Spring samples
			70	2.8	up to 24	11	Winter samples
	Springfield/Chicopee, MA	1987-88	49	0.3	up to 7.2	11	Spring samples
			50	0.08	up to 41	11	Winter samples
Hexachlorobenzene 118-74-1	*10 U.S. locations*	*1992-97*	*1,023*	*2 ng/m³*	*< 6 ng/m³*	*28*	*All nondetects*
	College Station, TX	*1979-80*	*16*	*0.21*	*0.14-0.39*	*1*	*Rural site*
	Denver, CO	*1980*	*?*	*0.24*	*-*	*8*	*Downtown, winter*
	Gulf Coast, TX	*1982*	*?*	*0.13*	*-*	*9*	
	Jacksonville, FL	*1987-88*	*60*	*0.46*	*< 0.5-13.0*	*11*	*Summer samples; includes 59 nondetects*
			72	*0.55*	*< 1.1*	*11*	*Spring samples; all nondetects*
			70	*0.50*	*< 1.0*	*11*	*Winter samples; all nondetects*
	Springfield/Chicopee, MA	*1987-88*	*49*	*0.95*	*< 1.9*	*11*	*Spring samples; all nondetects*
			50	*0.60*	*< 1.2*	*11*	*Winter samples; all nondetects*
							Detection limits estimated

Compound	Location	Period	Concentration	No.	Range	No.	Comments
Hexachlorobutadiene 87-68-3	180 U.S. locations	1988-99	0.357 µg/m³	12,914	< 0.001-48.4 µg/m³	28	Includes 12,126 nondetects
	8 U.S. urban to suburban locations	1978-81	0.11 (Median 0.03)	56	up to 3.71	71	Includes 8 nondetects
	Lima, OH	1990-91	0.55	81	< 1.1	21	All nondetects
1,2,3,4,5,6-Hexachloro-cyclohexane and isomers (e.g., Lindane 58-89-9)	10 U.S. locations	1992-97	2 ng/m³	1,023	< 6 ng/m³	28	All nondetects
	Baltimore, MD, Fresno, CA, Riverside, CA, SL City, UT	1967-68	–	437	up to 7.0	7,82	Includes 409 nondetects; urban sites
	Miami, FL	1973-74	0.7	14	0.2-2.5	82	Urban site
	College Station, TX	1979-80	0.23	14	0.01-1.60	1	Rural site
Hexachlorocyclo-pentadiene 77-47-4	10 U.S. locations	1992-97	0.002 µg/m³	1,023	< 0.006 µg/m³	28	All nondetects
Hexachloroethane 67-72-1	10 U.S. locations	1992-97	0.002 µg/m³	1,023	< 0.006 µg/m³	28	All nondetects
	26 U.S. urban to suburban locations	1976-78	0.01	69	< 0.02-0.20	71	Includes 63 nondetects
Hexamethylene diisocyanate 822-06-0	—	—	—	—	—	—	—
Hexamethyl-phosphoramide 680-31-9	—	—	—	—	—	—	—
Hexane 110-54-3	335 U.S. locations	1973-00	1.826 µg/m³	425,432	< 0.006-2,434 µg/m³	28	Includes 61,393 nondetects
	56 U.S. urban to suburban locations	1968-87	12.6 (Median 5.4)	1,590	up to 273	71	Includes 46 nondetects
Hydrazine 302-01-2	—	—	—	—	—	—	—

TABLE 4.1 (CONTINUED)
Ambient Air Concentrations of 188 Hazardous Air Pollutant (Chemicals shown in italics are high priority urban HAPs)

Compound and CAS Number	Locations	Year	N[a]	Concentration[b] Mean	Concentration[b] Range	Ref	Comments
Hydrogen chloride (HCl) 7647-01-0	Glendora, CA	1986 ca.	45	1.53 µg/m³	0.08-4.1 µg/m³	59	
	Los Angeles, CA (nine sites)	1987	–	–	1.2-2.8 (annual averages at individual sites)	60	
	Tucson, AZ	1981	–	–	Up to 3.0	61	
	Hampton, VA	1984	4	–	0.2-2.9	62	
	Claremont, CA	1985	16	–	0.2-1.5	63	
	Columbus, OH	1980	9	0.53	0.18-1.52	75	
Hydrogen fluoride (HF) 7664-39-3	6 U.S. locations	1992-97	1,567	0.049 µg/m³	0.03-2.7 µg/m³	28	Source impacted
	Unidentified U.S. location	ca. 1985	20	3.4	1.0 to 7.5	68	
Hydroquinone 123-31-9	—	—	—	—	—	—	—
Isophorone 78-59-1	10 U.S. locations	1992-97	906	5 ng/m³	< 1-396 ng/m³	28	Includes 794 nondetects
Maleic anhydride 108-31-6	—	—	—	—	—	—	—
Methanol 67-56-1	15 U.S. locations	1992-98	280	10.9 µg/m³	< 0.01-422 µg/m³	28	Includes 237 nondetects
	Boston, MA	1990-91	22	23.1 (Median 17.2)	9.5-62.3	20	
	Houston, TX	1990-91	22	22.0 (Median 17.6)	7.4-41.1	20	
	Lima, OH	1990-91	8	11.4	6.0-20.9	21	Two urban sites

Compound (CAS No.)	Location	Years	N	Mean (Median)	Range		Notes
Methoxychlor 72-43-5	Jacksonville, FL	1987-88	60	0.5 ng/m³	< 1.0 ng/m³	11	Summer samples; all nondetects
			72	2.1	< 4.2	11	Spring samples; all nondetects
			70	1.9	< 3.6-7.4	11	Winter samples
	Springfield/Chicopee, MA	1987-88	49	3.1	< 6.2	11	Spring samples; all nondetects
			50	2.3	< 4.5	11	Winter samples; all nondetects
							Detection limits estimated
Methyl bromide (Bromomethane) 74-83-9	337 U.S. locations	1983-00	29,716	1.164 µg/m³	< 0.019-4,889 µg/m³	28	Includes 26,237 nondetects
	35 U.S. urban to suburban locations	1975-86	353	8.83 (Median 0.67)	up to 149	71	Includes 33 nondetects
	Columbus, OH	1989	298	0.06	< 0.12	33	All nondetects
	Lima, OH	1990-91	81	0.20	< 0.39	21	All nondetects
	11 U.S. cities	1990	349	0.41	< 0.79-2.21	74	Includes 345 nondetects
Methyl chloride (Chloromethane) 74-87-3	261 U.S. locations	1982-00	19,836	1.538 µg/m³	< 0.021-350 µg/m³	28	Includes 3,931 nondetects
	24 U.S. urban to suburban locations	1975-86	706	1.55 (Median 1.37)	up to 12.0	71	Includes 9 nondetects
	Columbus, OH	1989	298	1.03	0.06-3.82	33	
	11 U.S. Cities	1990	349	0.25	< 0.46-5.1	74	Includes 347 nondetects
	Lima, OH	1990-91	81	0.44	< 0.2-2.2	21	Includes 43 nondetects
Methyl chloroform (1,1,1-Trichloroethane) 71-55-6	443 U.S. locations	1981-00	40,751	1.953 µg/m³	< 0.005-4,843 µg/m³	28	Includes 18,684 nondetects
	129 U.S. urban to suburban locations	1972-87	3,622	6.48 (Median 1.85)	up to 281	71	Includes 316 nondetects
	Columbus, OH	1989	298	4.00	0.78-46	33	
	11 U.S. cities	1990	349	7.73 (Median 2.11)	< 0.28-492	74	
	Lima, OH	1990-91	81	2.0	0.56-3.6	21	Includes 3 nondetects

TABLE 4.1 (CONTINUED)
Ambient Air Concentrations of 188 Hazardous Air Pollutant (Chemicals shown in italics are high priority urban HAPs)

Compound and CAS Number	Locations	Year	N[a]	Concentration[b] Mean	Concentration[b] Range	Ref	Comments
Methyl ethyl ketone (2-Butanone) 78-93-3	93 U.S. locations	1980-00	10,565	1.468 µg/m³	< 0.021-649 µg/m³	28	Includes 4,708 nondetects
	25 U.S. urban to suburban locations	1976-82	275	1.4	up to 36.9	71	Includes 246 nondetects
	Lima, OH	1990-91	8	4.0	0.6-15.3	21	
	Houston, TX	1990-91	22	15.3 (Median 2.97)	1.41-174	20	
	Boston, MA	1990-91	22	4.1 (Median 3.84)	< 0.6-14.7	20	Includes one nondetect
Methylhydrazine 60-34-4	—	—	—	—	—	—	—
Methyl iodide (Iodomethane) 74-88-4	6 U.S. locations	1983-85	77	0.173 µg/m³	< 0.0058-1.86 µg/m³	28	Includes 8 nondetects
	27 U.S. urban to suburban locations	1972-85	167	0.12 (Median 0.02)	up to 1.90	71	Includes 5 nondetects
Methyl iso-butyl ketone 108-10-1	53 U.S. locations	1992-00	1,232	1.913 µg/m³	< 0.01-320 µg/m³	28	Includes 993 nondetects
	Ringwood State Park, NJ	1983	?	—	0.52-5.87	72	Background VOC sampling
Methyl isocyanate 624-83-9	—	—	—	—	—	—	—
Methylmethacrylate 80-62-6	1 U.S. location	1992	7	8.70 µg/m³	7.12-10.4 µg/m³	28	

Compound / CAS No.	Location	Period	N	Mean concentration	Range	Ref.	Notes
Methyl tert-butyl ether 1634-04-4	148 U.S. locations	1995-00	12,895	3.35 µg/m³	< 0.036-849 µg/m³	28	Includes 6,959 nondetects
	Boston, MA	1990-91	22	0.41	< 0.7-1.7	20	Includes 21 nondetects
	Houston, TX	1990-91	22	1.4	< 0.7-9.5	20	Includes 14 nondetects
	Lima, OH, 2 sites	1990-91	8	0.35	< 0.7	21	All nondetects
	Houston, TX, 7 sites	1987-88	>100	0.35	< 0.7	70	All nondetects
4,4′-Methylenebis (2-chloro aniline) 101-14-4	—	—	—	—	—	—	—
Methylene chloride 75-09-2	481 U.S. locations	1983-00	54,405	2.40 µg/m³	< 0.035-2,812 µg/m³	28	Includes 32,234 nondetects
	65 U.S. urban to suburban locations	1976-87	2,238	3.91	up to 83	71	Includes 786 nondetects
	Columbus, OH	1989	298	1.70 (Median 1.19)	0.12-59	33	
	11 U.S. cities	1990	349	2.90	< 0.39-112	74	Includes 302 nondetects
	Lima, OH	1990-91	81	0.88	< 0.35-10.7	21	Includes 26 nondetects
4,4′-Methylenediphenyl diisocyanate 101-68-8	—	—	—	—	—	—	—
4,4′-Methylenedianiline 101-77-9	—	—	—	—	—	—	—
Naphthalene 91-20-3	42 U.S. locations	1992-98	511	468 ng/m³	< 1-18,900 ng/m³	28	Includes 395 nondetects
	5 U.S. urban to suburban locations	1974-82	7	530	up to 2,560	71	Includes 2 nondetects
	Boston, MA	1990-91	24	470 (Median 110)	142-1,165	53	
	Houston, TX	1990-91	21	436 (Median 448)	10-1,640	53	
	Chicago, IL	1991	16	530 (Median 326)	160-840	52	
	Kankakee, IL	1991	16	330	7.8-960	52	

TABLE 4.1 (CONTINUED)
Ambient Air Concentrations of 188 Hazardous Air Pollutant (Chemicals shown in italics are high priority urban HAPs)

Compound and CAS Number	Locations	Year	N^a	Concentration[b] Mean	Range	Ref	Comments
Nitrobenzene 98-95-3	10 U.S. locations	1992-97	1,023	0.003 µg/m³	< 0.008 µg/m³	28	All nondetects
	15 U.S. urban to suburban locations	1976-82	727	0.61	up to 61.4	71	Includes 442 nondetects
4-Nitrobiphenyl 92-93-3	Torrance, CA	1986	1	6.0 ng/m³ daytime 0.5 nighttime		6	3-nitrobiphenyl measured
	Boise, ID	1986-87	2	–	up to 0.06	22	Includes one nondetect
4-Nitrophenol 100-02-7	10 U.S. locations	1992-97	1,023	11 ng/m	< 2-21ng/m³	28	Includes 1,022 nondetects
	Boise, ID	1986-87	2	2.8	2.7-2.8	22	2-nitrophenol measured
2-Nitropropane 79-46-9	—	—	—	—	—	—	—
N-nitroso-N-methylurea 684-93-5	—	—	—	—	—	—	—
N-nitrosodimethylamine 62-75-9	10 U.S. locations	1992-97	1,023	2 ng/m³	< 1-58 ng/m³	28	Includes 1,020 nondetects
N-Nitrosomorpholine 59-89-2	—	—	—	—	—	—	—
Parathion 56-38-2	Orlando, FL	1967-68	14	8.5 ng/m³	up to 25.4	7	Includes 7 nondetects; rural site; 14 representative samples shown of 99 samples (of all samples: 62 NDs; max=465 ng/m³). Parathion was detected in only one of a total of 5 rural sites in this study.

Chemical / CAS	Location	Year	No. of samples	Concentration	Range	Ref	Comments
Pentachloronitrobenzene 82-68-8	—	—	—	—	—	—	—
Pentachlorophenol 87-86-5	11 U.S. locations	1992-97	1,033	4 ng/m³	< 1-22 ng/m³	28	Includes 1,032 nondetects
	Columbia, SC	1989	2	0.92	0.91-0.92	16	
Phenol 108-95-2	21 U.S. locations	1992-97	1,027	0.155 µg/m³	< 0.001-57.7 µg/m³	28	Includes 851 nondetects
	3 U.S. urban locations	1974-82	4	< 0.03	up to 0.12	71	Includes 3 nondetects
p-Phenylenediamine 106-50-3	—	—	—	—	—	—	—
Phosgene 75-44-5	8 U.S. urban to suburban locations	1976-78	10	0.08 µg/m³ (Median 0.08)	0.05-0.19 µg/m³	71	
	Staten Island, NY	1983	8	0.45 (Median 0.39)	0.29-0.78	76	
Phosphine 7803-51-2	—	—	—	—	—	—	—
Phosphorus (elemental) 7723-14-0	—	—	—	—	—	—	—
Phthalic anhydride 85-44-9	Unspecified location, probably Northeast U.S.	1983	>10	--	< 6.2 µg/m³	23	Upwind of waste sites, and in residential areas; number of samples estimated
Polychlorinated biphenyls 1336-36-3	10 U.S. Cities	1973-78	Many	5 ng/m³	0.5-36 ng/m³	27	Review of other studies
	Great Lakes Region	Up to 1991	Many	1.0	--	29	Review of other studies
	South Carolina	1976-79	37	3.1	--	30	
	Madison, WI	1977	–	7.7	7.3-8.1	31	
		1979	–	3.0	0.8-5.0		
	Minneapolis, MN	1978	26	7.0	1.3-20	32	
		1979	10	7.1	4.3-9.1		
	Columbia, SC	1977-79	–	4.7	--	8	
	Denver, CO	1980	–	2.3	--	8	

TABLE 4.1 (CONTINUED)
Ambient Air Concentrations of 188 Hazardous Air Pollutant (Chemicals shown in italics are high priority urban HAPs)

Compound and CAS Number	Locations	Year	N[a]	Concentration[b] Mean	Concentration[b] Range	Ref	Comments
1,3-propane sultone 1120-71-4	—	—	—	—	—	—	—
β-Propiolactone 57-57-8	—	—	—	—	—	—	—
Propionaldehyde 123-38-6	127 U.S. locations	1980-00	18,625	0.757 µg/m³	< 0.007-78.2 µg/m³	28	Includes 12,393 nondefects
	17 U.S. urban to suburban locations	1974-80	94	9.98	up to 77.4	71	Includes 51 nondetects
Propoxur (Baygon) 114-26-1	Jacksonville, FL	1987-88	47	10.2 ng/m³	up to 206 ng/m³	11	Summer samples
			72	0.8	up to 20		Spring samples
			70	2.5	up to 65		Winter samples
	Springfield/Chicopee, MA	1987-88	49	0.8	up to 33	11	Spring samples
			50	0.07	up to 23		Winter samples
Propylene dichloride (1,2-Dichloropropane) 78-87-5	*336 U.S. locations*	*1983-00*	*30,842*	*0.503 µg/m³*	*< 0.009-118 µg/m³*	*28*	*Includes 29,696 nondetects*
	14 U.S. urban to suburban locations	*1979-85*	*708*	*0.75 (Median 0.09)*	*up to 59.6*	*71*	*Includes 146 nondetects*
	Columbus, OH	*1989*	*298*	*0.07*	*< 0.14*	*33*	*All nondetects*
	11 U.S. cities	*1990*	*349*	*0.56*	*< 0.19-12.5*	*74*	*Includes 281 nondetects*
	Lima, OH	*1990-91*	*81*	*0.24*	*< 0.47*	*21*	*All nondetects*
Propylene oxide 75-56-9	2 U.S. locations	1990-91	44	57.4 µg/m³	9.7-224 µg/m³	28	Probably near source

Chemical / CAS	Location	Year	No. samples	Concentration	Range	Ref.	Comments
1,2-Propyleneimine (2-methyl aziridine) 75-55-8						—	—
Quinoline 91-22-5	2 U.S. urban locations	1982	3	< 0.34 µg/m³	up to 1.0 µg/m³	71	Includes 2 nondetects
Quinone 106-51-4						—	—
Styrene 100-42-5	554 U.S. locations	1981-00	429,901	0.468 µg/m³	< 0.001-913 µg/m³	28	Includes 196,813 nondetects
	18 U.S. urban to suburban locations	1968-86	1,118	3.7 (Median 0.82)	up to 279	71	Includes 279 nondetects
	Columbus, OH	1989	298	0.19	< 0.13-3.55	33	Includes 265 nondetects
	Atlanta, GA	1990	4,620	0.58 (Median 0.34)	< 0.05-35.1	73	
	Lima, OH	1990-91	81	0.22	< 0.43	21	All nondetects
Styrene Oxide 96-09-3						—	—
2,3,7,8 Tetrachlorodibenzo-p-dioxin 1746-01-6	Columbus, Akron, Waldo, OH	1987	6	0.12 pg/m³	< 0.012 to < 0.82 pg/m³	17	All nondetects
	Bloomington, IN	1986-88	95	0.048 pg/m³	—	18	Sum of isomers (2,3,7,8+ 2,3,4,8+ 2,3,4,6); Geometric mean of vapor + particle concentrations
	South Coast Air Basin, CA	1987	33	0.012 pg/m³	—	10	Includes 30 nondetects Avg. DL=0.022 pg/m³; 8 sites
	Niagara Falls, NY	1986-87	6	0.007	< 0.014	84	All nondetects
	3 New York cities	1988	6	0.007	< 0.014	85	All nondetects
	Columbus, OH, 3 sites	1994-95	14	0.005	—	86-88	Urban sites
			2	0.019	—		Urban-source impacted
			3	0.003	—		Upwind rural
	Mohawk Mountain, CT	1993-94	4	< 0.001	up to 0.001	89	Near major road
	Phoenix, AZ	1994	4	0.008	—	90	

TABLE 4.1 (CONTINUED)
Ambient Air Concentrations of 188 Hazardous Air Pollutants (Chemicals shown in italics are high priority urban HAPs)

Compound and CAS Number	Locations	Year	N[a]	Concentration[b] Mean	Range	Ref	Comments
1,1,2-Tetrachloroethane *79-34-5*	*232 U.S. locations*	*1983-99*	*13,457*	*0.27 µg/m³*	*< 0.014-38.6 µg/m³*	*28*	*Includes 12,803 nondetects*
	39 U.S. urban to suburban locations	*1976-86*	*876*	*0.49*	*up to 14*	*71*	*Includes 592 nondetects*
	11 U.S. cities	*1990*	*349*	*0.14 (Median < 0.07)*	*< 0.07-7.3*	*74*	*Includes 322 nondetects*
	Lima, OH	*1990-91*	*81*	*0.35*	*< 0.70*	*21*	*All nondetects*
Tetrachloroethylene *(Perchloroethylene)* *127-18-4*	*482 U.S. locations*	*1981-00*	*46,899*	*1.651 µg/m³*	*< 0.034-6020 µg/m³*	*28*	*Includes 29,597 nondetects*
	120 U.S. urban to suburban locations	*1972-87*	*4,165*	*4.84 (Median 2.07)*	*up to 296*	*71*	*Includes 372 nondetects*
	Columbus, OH	*1989*	*298*	*1.59*	*0.21-40*	*33*	
	11 U.S. cities	*1990*	*349*	*6.15 (Median 2.14)*	*< 0.48-104*	*74*	*Includes 16 nondetects*
Titanium tetrachloride 7550-45-0	—	—	—	—	—	—	—
Toluene 108-88-3	835 U.S. locations	1973-00	468,189	5.75 µg/m³	< 0.005-3345 µg/m³	28	Includes 16,834 nondetects
	117 U.S. urban to suburban locations	1968-87	4,025	24.2 (Median 10.9)	up to 965	71	Includes 122 nondetects
	Columbus, OH	1989	298	5.17	0.11-20.6	33	
	Atlanta, GA	1990	4,620	9.9 (Median 6.2)	0.38-561	73	Corrected for minor contribution of coeluting 2-methylheptane
	11 U.S. cities	1990	349	19.9 (Median 7.4)	0.23-750	74	
	Lima, OH	1990-91	81	5.78	1.69-16.5	21	

Compound CAS	Location	Year	No.	Mean	Range	Ref	Comments
Toluene 2,4-diamine 95-80-7	—	—	—	—	—	—	
2,4-Toluene diisocyanate 584-84-9	—	—	—	—	—	—	
o-Toluidine 95-53-4	—	—	—	—	—	—	
Toxaphene 8001-35-2	College Station, TX	1979-80	3	1.8 ng/m³	1.0–2.5 ng/m³	1	
	Columbia, SC	1976-79	?	13.1	—	12	
	Gulf Coast, TX	1982	?	0.6	—	9	
	Wye River, MD	1982	8	98	11–376	24	Summer time, weekly, day-time measurements
	Greenville, MS	1981	15	7.0	1.8–22	25	Summer and fall
	St. Louis, MO		15	1.3	0.37–5.2	25	
	Bridgman, MI		14	0.36	0.01–1.2	25	Strong seasonal variation, summer peak
	Columbia, SC	1978-79	>30	13.1	<1–66	8,26	
1,2,4-Trichlorobenzene 120-82-1	161 U.S. locations	1986-99	11,838	0.281 µg/m³	<0.001–24.7 µg/m³	28	Includes 10,794 nondetects
	1 U.S. urban location	1986	18	1.28 (Median 0.76)	0–3.78	71	Includes 2 nondetects
	Lima, OH	1990-91	81	0.38	<0.76	21	All nondetects
1,1,2-Trichloroethane 79-00-5	361 U.S. locations	1983-00	34,990	0.220 µg/m³	<0.055–27.6 µg/m³	28	Includes 34,312 nondetects
	41 U.S. urban to suburban locations	1976-86	855	3.71	up to 315	71	Includes 541 nondetects
	11 U.S. cities	1990	349	0.33 (Median <0.22)	<0.22–15.9	74	Includes 290 nondetects
	Lima, OH	1990-91	81	0.28	<0.56	21	All nondetects

TABLE 4.1 (CONTINUED)
Ambient Air Concentrations of 188 Hazardous Air Pollutants (Chemicals shown in italics are high priority urban HAPs)

Compound and CAS Number	Locations	Year	N[a]	Concentration[b] Mean	Range	Ref	Comments
Trichloroethylene 79-01-6	*426 U.S. locations*	*1981-00*	*48,411*	*0.96 µg/m³*	*< 0.011-321 µg/m³*	*28*	*Includes 35,235 nondetects*
	111 U.S. urban to suburban locations	*1973-87*	*3,539*	*2.08 (Median 0.66)*	*up to 376*	*71*	*Includes 812 nondetects*
	Columbus, OH	*1989*	*298*	*0.42*	*< 0.16-17.8*	*33*	*Includes 247 nondetects*
	11 U.S. cities	*1990*	*349*	*2.63 (Median < 0.05)*	*< 0.05-162*	*74*	*Includes 226 nondetects*
	Lima, OH	*1990-91*	*81*	*0.71*	*< 0.55-17.6*	*21*	*Includes 64 nondetects*
2,4,5-Trichlorophenol 95-95-4	10 U.S. locations	1992-97	1,023	4 ng/m³	< 21 ng/m³	28	All nondetects
	Portland, OR	1984	7	0.15	up to 0.31	5	(Sum of 2,4,5 and 2,4,6-trichlorophenol)
	Columbia, SC	1989	2	0.06	< 0.07-0.08	17	
2,4,6-Trichlorophenol 88-06-2	10 U.S. locations	1992-97	1,023	4 ng/m³	< 21 ng/m³	28	All nondetects
	Portland, OR	1984	7	0.15	up to 0.31	5	(Sum of 2,4,5 and 2,4,6-trichlorophenol)
	Columbia, SC	1989	2	0.25	0.20-0.29	17	
Triethylamine 121-44-8	Unspecified location, Northeast U.S. (suspected)	1983	2	2.1 µg/m³	< 4.2 µg/m³	23	Both nondetects
Trifluralin 1582-09-8	Egbert, ON, Canada	1988-89	143	0.27 ng/m³	up to 3.4 ng/m³	14	Includes 68 nondetects
2,2,4-Trimethylpentane 540-84-1	200 U.S. locations	1973-00	416,455	1.35 µg/m³	< 0.006-583 µg/m³	28	Includes 61,394 nondetects
	23 U.S. urban to suburban locations	1971-86	501	22.5 (Median 7.1)	up to 230	71	Includes 61 nondetects

Compound / Location	Period	Concentration	N	Range	N	Comments
Vinyl acetate 108-05-4						
7 U.S. locations	1993-99	2.00 µg/m³	86	< 0.352-17.6 µg/m³	28	Includes 57 nondetects
Lima, OH	1990-91	3.5	8	< 7	21	Two urban sites, all nondetects
Houston, TX	1990-91	3.5	22	< 7	20	All nondetects
Boston, MA	1990-91	3.5	22	< 7	20	All nondetects
Vinyl Bromide 593-60-2						
—	—	—	—	—	—	—
Vinyl chloride 75-01-4						
365 U.S. locations	*1986-00*	*0.408 µg/m³*	*40,094*	*< 0.026-125 µg/m³*	*28*	*Includes 39,032 nondetects*
Columbus, OH	*1989*	*0.04*	*298*	*< 0.08*	*33*	*All nondetects*
11 U.S. cities	*1990*	*1.95*	*349*	*< 0.5-202*	*74*	*Includes 339 nondetects*
Lima, OH	*1990-91*	*0.13*	*81*	*< 0.26*	*21*	*All nondetects*
Vinylidene chloride (1,1-Dichloroethylene) 75-35-4						
239 U.S. locations	1981-00	0.308 µg/m³	24,299	< 0.036-122 µg/m³	28	Includes 22,543 nondetects
Columbus, OH	1989	0.06	298	< 0.12	33	All nondetects
Lima, OH	1990-91	0.20	81	< 0.40	21	All nondetects
Xylenes (mixed) 1330-20-7						
41 U.S. locations	1990-95	12.157 µg/m³	300	< 0.434-326 µg/m³	28	Includes 94 nondetects; See entries for o,m, p-isomers.
o-Xylene 95-47-6						
669 U.S. locations	1973-00	1.247 µg/m³	461,571	< 0.005-3,854 µg/m³	28	Includes 89,229 nondetects
101 U.S. urban to suburban locations	1973-86	8.5 (Median 3.2)	3,543	up to 391	71	Includes 132 nondetects
Columbus, OH	1989	1.37	298	0.13-5.38	33	
Atlanta, GA	1990	2.68 (Median 1.67)	4,620	< 0.05-64.1	73	Upper limit; includes minor contribution from coeluting n-nonane
Lima, OH	1990-91	1.5	81	< 0.44-4.1	21	Includes 3 nondetects

TABLE 4.1 (CONTINUED)
Ambient Air Concentrations of 188 Hazardous Air Pollutants (Chemicals shown in italics are high priority urban HAPs)

Compound and CAS Number	Locations	Year	N^a	Concentration[b] Mean	Range	Ref	Comments
m-Xylene 108-38-3	146 U.S. locations	1973-99	16,520	3.508 µg/m³	< 0.027-283 µg/m³	28	Includes 3,634 nondetects
	95 U.S. urban to suburban locations	1968-85	3,732	11.0 (Median 6.7)	up to 335	71	Includes 399 nondetects
	Columbus, OH	1989	298	3.62	0.13-17.1	33	Sum of m+p isomers
	Atlanta, GA	1990	4,620	5.54 (Median 3.47)	< 0.06-127	73	Sum of m+p isomers
	Lima, OH	1990-91	81	3.9	0.44-11.3	21	Sum of m+p isomers
p-Xylene 106-42-3	169 U.S. locations	1973-99	15,302	2.570 µg/m³	< 0.027-468 µg/m³	28	Includes 3,095 nondetects
	102 U.S. urban to suburban locations	1968-87	3,597	14.8 (Median 6.9)	up to 377	71	Includes 106 nondetects
Antimony compounds	28 U.S. locations	1993-98	5,927	2 ng/m³	4-220 ng/m³	28	Includes 5,821 nondetects
	Ohio River Valley	1980-81	1,204	1.7	< 3.3	43	All nondetects
Arsenic Compounds	*32 U.S. locations*	*1990-98*	*8,431*	*1 ng/m³*	*2-50 ng/m³*	*28*	*Includes 8,129 nondetects*
	Columbus, OH	*1989*	*353*	*2.5*	*< 5*	*33*	*All nondetects*
	Great Lakes Region	*Up to 1991*	*–*	*0.5*	*–*	*29*	*Review of other studies*
	New York City	*1977-81*	*–*	*–*	*2.6–10.9*	*40*	
	South Coast Air Basin	*1985*	*343*	*2.6*	*1.8–7.0 (individual site means)*	*41*	
	Ohio River Valley	*1980-81*	*1,204*	*< 1.3*	*< 1.3–12*	*43*	
Beryllium Compounds	*30 U.S. locations*	*1990-00*	*5,218*	*0.3 ng/m³*	*< 0.001-18 ng/m³*	*28*	*Includes 4,961 nondetects*
	30 U.S. cities	*1978*	*–*	*0.3*	*0.1-0.5*	*37*	
	South Coast Air Basin, CA	*1985*	*354*	*0.05*	*< 0.1*	*41*	*All nondetects*
	17 U.S. cities	*1978*	*–*	*–*	*0.1-0.5*	*58*	

Compound	Location	Year	n	Concentration	Range	Ref.	Notes
Cadmium Compounds	22 U.S. locations	1990-92	2,093	3 ng/m³	2.5-17 ng/m³	28	
	U.S. urban areas	1979	–	(Median < 6)	< 6-200	36	Review of other studies
	Great Lakes region	Up to 1991	–	0.5	–	27	
	South Coast Air Basin, CA	1985	349	1.2	0.3-4.1 (individual site means)	41	
Chromium Compounds	118 U.S. locations	1988-99	70,917	1 ng/m³	< 0.4-95 ng/m³	28	Includes 54,176 nondetects
	Columbus, OH	1989	353	< 5	< 5-30	33	
	Great Lakes Region	Up to 1991	–	2	–	29	Review of other studies
	Lima, OH	1990-91	112	0.5	< 1	21	All nondetects
	South Coast Air Basin, CA	1985	193	5.1	Individual site means 3.1-10	41	
	U.S. urban areas	1984-87	–	7	–	42	
	Ohio River Valley	1980-81	1,204	3.3	< 6.6	43	All nondetects
	186 U.S. urban sites	1978	–	–	< 5 to 120	35	
Cobalt Compounds	32 U.S. locations	1990-98	8,431	3 ng/m³	< 6-12 ng/m³	28	Includes 8,430 nondetects
	Columbus, OH	1989	353	2.5	< 5	33	All nondetects
	28 U.S. stations	1977	750	0.15	< 0.3	34	All nondetects
	Chicago, IL	1977	750	–	0.3-23	34	
	South Coast Air Basin, CA	1985	39	1.1	1.0-1.1	41	
	Ohio River Valley	1980-81	1,204	1.2	< 2.3	43	All nondetects
Coke Oven Emissions	—	—	—	—	—	—	
Cyanide compounds	—	—	—	—	—	—	
Glycol Ethers	—	—	—	—	—	—	

TABLE 4.1 (CONTINUED)
Ambient Air Concentrations of 188 Hazardous Air Pollutants (Chemicals shown in italics are high priority urban HAPs)

Compound and CAS Number	Locations	Year	N[a]	Concentration[b] Mean	Concentration[b] Range	Ref	Comments
Lead Compounds	*117 U.S. locations*	*1988-99*	*69,606*	*2 ng/m³*	*< 0.06-320 ng/m³*	*28*	*Includes 9,396 nondetects*
	Columbus, OH	1989	353	10	< 5–50	33	
	Great Lakes Region	Up to 1991	–	8	–	29	Review of other studies
	Lima, OH	1990-91	112	6	0.4-20	21	
	South Coast Air Basin, CA	1985	50	226	Individual site means 180-280	41	2 sites
	U.S. urban areas	1984-87	–	124	Individual site means	42	Fine = < 2.5 µm
	Ohio River Valley	1980-81	1,204	52 Fine 9 Coarse	45-56 8-9 Max individual sample: 213	43	Coarse = 2.5 to 10 µm
Manganese Compounds	*118 U.S. locations*	*1988-99*	*70,796*	*1 ng/m³*	*< 0.4-113 ng/m³*	*28*	*Includes 36,212 nondetects*
	Columbus, OH	1989	353	2.5	< 5–20	33	
	U.S. urban sites	1982	–	24	–	3	
	Lima, OH	1990-91	112	3	0.5-10.4	21	
	South Coast Air Basin, CA	1985	199	31	Individual site means 25-44	41	
	U.S. urban areas	1984-87	–	45	–	42	
	Ohio River Valley	1980-81	1204	6	< 3.3 to 9 (individual site means)	43	

Mercury Compounds	Perch River, NY	1991-92	230	2.5 ng/m³	0.5-7 ng/m³	Vapor phase	46
			243	0.062	up to 0.75	Fine particles (PM$_{2.5}$)	46
			111	0.025	up to 0.11	Coarse particles (PM$_{10}$-PM$_{2.5}$)	46
	Coastal Connecticut	1998	15,280	1.85	up to 10.1	Total gaseous mercury	45
	Rural Wisconsin	1988-89	36	1.57	1.00-2.45	Total gaseous mercury (>99.3% elemental Hg, < 0.7% monomethyl mercury)	83
	Avery Point, CT	1989	15	0.022	0.007-0.062	Particle phase	83
		1988-89	22	0.062	0.005-0.18	Particle phase	83
	Broward County, FL (3 sites)	1993	–	1.8-3.3 (Site means)	0.9-4.6	Vapor phase	91
		1993	–	0.034-0.051 (Site means)	0.013-0.120	Particle phase	91
	Great Lakes Region	1992	–	2.0-8.7	1.3-62.7	Vapor phase	29
	(total of 6 sites)			0.019-0.30 (5 rural sites)	0.009-1.23	Particle phase	29
		1992	–	>40.8 (One Detroit site)	up to >74	Vapor phase	29
				0.34	up to 1.09	Particle phase	29
	Oak Ridge, TN	1988-89	84	5.5	0.8-16	Vapor phase	47
	Nine U.S. cities	1974	9	9.9	5.0-14.4	Vapor phase	69
Fine Mineral Fibers	U.S. urban areas	—	—	Concentrations probably roughly equal to those of asbestos fibers		See entry for asbestos	54

TABLE 4.1 (CONTINUED)
Ambient Air Concentrations of 188 Hazardous Air Pollutants (Chemicals shown in italics are high priority urban HAPs)

Compound and CAS Number	Locations	Year	N[a]	Mean	Concentration[b] Range	Ref	Comments
Nickel Compounds	*118 U.S. locations*	*1988-99*	*70,798*	*0.3 ng/m³*	*< 0.05-55 ng/m³*	*28*	*Includes 51,616 nondetects*
	Columbus, OH	*1989*	*353*	*2.5*	*< 5*	*33*	*All nondetects*
	South Coast Air Basin, CA	*1985*	*199*	*7.6*	*Individual site means*	*41*	
					5.4-8.7		
	U.S. urban areas	*1984-87*	*–*	*7*	*–*	*42*	
	Ohio River Valley	*1980-81*	*1,204*	*0.9*	*< 1.7 to 14*	*43*	
	Urban U.S. sites	*1970-80*	*–*	*9-24*	*9-640*	*38*	
				(site means)			
Polycyclic Organic Matter (see text)	*Los Angeles, CA (7 sites)*	*1981-82*	*24*	*6.4 ng/m³*	*0.3-24.5 ng/m³*	*48*	*7 out of 8 carcinogenic PAHs measured (not dibenzo (a,h)-anthracene) Range of seasonal means is shown*
	Newark, NJ	*1981-83*	*156*	*7.6*	*1.7-15.9*	*49*	*7 out of 8. Range of seasonal means is shown*
	Camden, NJ	*1981-83*	*156*	*5.3*	*1.1-9.7*	*49*	
	Elizabeth, NJ	*1981-83*	*156*	*7.3*	*1.4-14.8*	*49*	
	Columbus, OH	*1984-88*	*33*	*7.4*	*3.4-16.2*	*50*	*7 out of 8*
	San Gabriel Valley, CA (5 sites)	*1989*	*29*	*5.4*	*1.3-20.8*	*51*	
	Chicago, IL	*1991*	*16*	*29.8*	*2.9-91*	*52*	
	Kankakee, IL	*1991*	*16*	*2.1*	*0.3-17.9*	*52*	
	Boston, MA	*1990-91*	*24*	*8.0*	*1.8-24.8*	*53*	
	Houston, TX	*1990-91*	*22*	*3.0*	*0.4-10.3*	*53*	
	Minneapolis, MN and Salt Lake City, UT	*1988-89*	*22*	*8.9*	*< 2.43*	*67*	*Data from two cities pooled; winter data only*

Compound	Location	Date	N	Mean	Range	Ref.	Comments
Radionuclides (including Radon)	Continental U.S.	—	—	70-200 pCi/m³ (Radon only)	(Total natural radiation dose from all sources of ~500-1000 mrem/yr)	64-66	Excludes indoor radon. Radon contributes ~90% of radiation dose, via inhalation. Concentration range shown is in pico-Curies per m³ (pCi/m³).
Selenium Compounds	86 U.S. locations	1988-99	62,490	0.4 ng/m³	< 0.03-45 ng/m³	28	Includes 23,402 nondetects
	Columbus, OH	1989	353	2.5	< 5	33	All nondetects
	Lima, OH	1990-91	112	0.5	< 1	21	All nondetects
	Ohio River Valley	1980-81	1204	2	< 2-8	43	
	Cambridge, MA	1965	7	0.9	–	39	

Notes:

a: N = number of total measurements.

b: Units of concentration given in mass/volume units; µg/m³ = micrograms/m³; ng/m³ = nanograms/m³; pg/m³ = picograms/m³. Mean calculated using one half of detection limit for non-detects (multi-site or multi-year studies may have more than one reported detection limit).

Table References

1. Atlas, E. and Giam, C.S., Ambient concentrations and precipitation scavenging of atmospheric organic pollutants, *Water, Air, Soil Pollution*, 38(1-2), 19, 1988.
2. Radian Corporation, Urban air toxics monitoring program, 1990. EPA-450/4-91-024, U.S. Environmental Protection Agency, Research Triangle Park, NC, 285 pp. (NTIS No. PB92-110022), 1991.
3. Rogozen, M.B. et al., Evaluation of potential toxic air contaminants. Phase I. Final Report, ARB-R-88-333, California Air Resources Board, Sacramento, California, 582 pp. (NTIS No. PB88-183330), 1987.
4. Shields, H.C. and Weschler, C.S., Analysis of ambient concentrations of organic vapors with a passive sampler, *J. Air Poll. Control Assoc.*, 37(9), 1039, 1987.
5. Leuenberger, C., Ligocki, M.P., and Pankow, J.F., Trace organic compounds in rain. 4. Identities, concentrations, and scavenging mechanisms for phenols in urban air and rain. *Environ. Sci. Technol.*, 19(11):1053, 1985.
6. Arey, J. et al., Polycyclic aromatic hydrocarbon and nitroarene concentrations in ambient air during a wintertime high-NOₓ episode in the Los Angeles basin, *Atmos. Environ.*, 21(6) 1437, 1987.
7. Stanley, C.W et al., Measurement of atmospheric levels of pesticides, *Environ. Sci. Technol.*, 5(5), 430, 1971.
8. Billings, W.N. and Bidleman, T.F., High volume collection of chlorinated hydrocarbons in urban air using three solid adsorbents, *Atmos. Environ.*, 17(2), 383, 1983.
9. Chang, L.W., Atlas, E. and Giam, C.S., Chromatographic separation and analysis of chlorinated hydrocarbons and phthalic acid esters from ambient air samples, *Int. J. Environ. Anal. Chem.*, 19: 145, 1985.
10. Hunt, G.T. and Maisel, B.E., Atmospheric concentrations of PCDDs/PCDFs in Southern California, *J. Air Waste Manage. Assoc.*, 42, 672, 1992.

TABLE 4.1 (CONTINUED)
Ambient Air Concentrations of 188 Hazardous Air Pollutants

11. Immerman, F.W. and Schaum, J.L., Nonoccupational pesticide exposure study (NOPES), EPA-600/3-90-003, U.S. Environmental Protection Agency, Research Triangle Park, NC, 262 pp. (NTIS No. PB90-152224), 1990.

12. Bidleman, T.F. and Christenson, E.J., Atmospheric removal processes for high molecular weight organochlorines, *J. Geophys. Res.*, 84: 7857, 1979.

13. Syracuse Research Corporation, Criteria document for 2,4-dichlorophenoxyacetic acid, SRC TR-81-856, U.S. Environmental Protection Agency, Cincinnati, OH, 1981.

14. Hoff, R.M., Muir, D.C.G. and Grift, N.P., Annual cycle of polychlorinated biphenyls and organohalogen pesticides in air in southern Ontario 1. Air concentration data, *Environ. Sci. Technol.*, 26(2), 266, 1992.

15. Bove, J., Dalvent, P. and Kukreja, V.P., Airborne di-butyl and di-(2-ethylhexyl) phthalate at three New York City air sampling stations, *Int. J. Environ. Anal. Chem.*, 5: 189, 1978.

16. McConnell, L.L. et al., Development of a collection method for chlorophenolic compounds in air. In: *Proc. 1989 EPA/A&WMA Int. Symp. Measurement of Toxic and Related Air Pollutants*, A&WMA Publication VIP-13, EPA Report No. 600/9-89-060, Pittsburgh, PA, 623, 1989.

17. Edgerton, S.A. et al., Ambient air concentrations of polychlorinated dibenzo-p-dioxins and dibenzofurans in Ohio: sources and health risk assessment, *Chemosphere*, 18, 1713, 1989.

18. Eitzer, B.D. and Hites, R.A., Polychlorinated dibenzo-p-dioxins and dibenzofurans in the ambient atmosphere of Bloomington, IN, *Environ. Sci. Technol.*, 23(11), 1389, 1989.

19. Hunt, W.F., Faoro, R.B. and Freas, W., Report on the interim database for state and local air toxic volatile organic chemical measurements, EPA 450/4-86-012, U.S. Environmental Protection Agency, Research Triangle Park, NC, 1986.

20. Kelly, T.J., Callahan, P.J. and Pleil, J., Method development and field measurements for polar VOCs in ambient air, *Environ. Sci. Technol.*, 27, 1146, 1993.

21. Kelly, T.J. et al., Results of air pollutant measurements in Allen County, OH, Vol. I, Final Report to the Allen County Board of Commissioners, Battelle, Columbus, OH, October 1991.

22. Nishioka, M.G. and Lewtas, J., Quantification of nitro- and hydroxylated nitro-aromatic/polycyclic aromatic hydrocarbons in selected ambient air daytime Winter samples, *Atmos. Environ.*, 26A, 2077, 1982.

23. Clay, P.F. and Spittler, T.M., Determination of airborne volatile nitrogen compounds using four independent techniques, Natl. Conf. Manage. Uncontrolled Hazard Waste Sites. Hazard. Mater. Control Res. Inst.: Silver Spring, MD, 100, 1983.

24. Glotfelty, D.E. et al., Regional atmospheric transport and deposition of pesticides in Maryland. In: D.A. Kurtz (Ed.), Long Range Transport of Pesticides. Lewis: Chelsea, MI, 199, 1990.

25. Rice, C.P., Samson, P.J. and Noguchi, G.E. Atmospheric transport of toxaphene to Lake Michigan, *Environ. Sci. Technol.*, 20, 1109, 1986.

26. Billings, W.N. and Bidleman, T.F., Field comparison of polyurethane foam and Tenax-GC resin for high-volume air sampling of chlorinated hydrocarbons, *Environ. Sci. Technology*, 14, 679, 1980.

27. Environmental Profiles and Hazard Indices for Constituents of Municipal Sludge: Polychlorinated Biphenyls, Report to U.S. EPA, Office of Water Regulations and Standards, prepared by Battelle, June 1985.

28. Rosenbaum, A.S., Stiefer, P.S. and Iwamiya, R.K., Air Toxics Data Archive and AIRS Combined Data Set: Data Base Descriptions, prepared for U.S. Environmental Protection Agency, Office of Policy, Planning, and Evaluation by Systems Applications International, SYSAPP-99/25, 1999.

29. Keeler, G., Hoyer, M. and Lamborg, C., Measurements of atmospheric mercury in the Great Lakes Basin, in *Mercury Pollution: Integration and Synthesis*, Watras, C.J. and Huckabee, J.W., Eds., Lewis, Boca Raton, FL, 231, 1994.

31. Doskey, P.V. and Andren, A.W., Concentrations of airborne PCBs over Lake Michigan, *J. Great Lakes Res.*, 7, 15, 1981.

32. Eisenreich, S.J., Looney, B.B. and Hollod, G.J., PCBs in the Lake Superior atmosphere 1978-80, in Physical Behavior of PCB's in the Great Lakes, D. MacKay, S. Paterson, S.J. Eisenreich, and M.S. Simmons, Eds., Ann Arbor Science, Ann Arbor, MI, 115, 1983.

33. Spicer, C.W. et. al, Variability and source attribution of hazardous urban air pollutants: Columbus field study, Final Report to U.S. Environmental Protection Agency, Contract No. 68-D-80082, Battelle, Columbus, OH, June 1992. Summarized in Spicer, C.W. et al., The variability of hazardous air pollutants in an urban area, *Atmos. Environ., Part B: Urban Atmospheres*, 30, 3443, 1996.

34. Environmental Profiles and Hazard Indices for Constituents of Municipal Sludge: Cobalt, Report to U.S. EPA, Office of Water Regulations and Standards, prepared by Battelle, June 1985.

35. Environmental Profiles and Hazard Indices for Constituents of Municipal Sludge: Chromium, Report to U.S. EPA, Office of Water Regulations and Standards, prepared by Battelle, June 1985.

36. Environmental Profiles and Hazard Indices for Constituents of Municipal Sludge: Cadmium, Report to U.S. EPA, Office of Water Regulations and Standards, prepared by Battelle, June 1985.

37. Environmental Profiles and Hazard Indices for Constituents of Municipal Sludge: Beryllium, Report to U.S. EPA, Office of Water Regulations and Standards, prepared by Battelle, June 1985.

38. Environmental Profiles and Hazard Indices for Constituents of Municipal Sludge: Nickel, Report to U.S. EPA, Office of Water Regulations and Standards, prepared by Battelle, June 1985.

39. Environmental Profiles and Hazard Indices for Constituents of Municipal Sludge: Selenium, Report to U.S. EPA, Office of Water Regulations and Standards, prepared by Battelle, June 1985.

40. Environmental Profiles and Hazard Indices for Constituents of Municipal Sludge: Arsenic, Report to U.S. EPA, Office of Water Regulations and Standards, prepared by Battelle, June 1985.

41. Multiple Air Toxics Exposure Study Working Paper No. 3, South Coast Air Basin, U.S. EPA report number EPA-450/4-88-013, prepared by South Coast Air Quality Management District and Systems Applications, Inc., November 1988.

42. Data from urban sites in the National Particulate Network, for the years 1984-87, cited in Ashland Air Toxics Survey, October 1988-February 1989, prepared by U.S. EPA Region IV, Atlanta, GA, June 1989.

43. Shaw, Jr., R.W. and Paur, R.J., Composition of aerosol particles collected at rural sites in the Ohio River Valley, *Atmos. Environ.*, 17, 2031, 1983.

44. World Health Organization, Mercury-Environmental Aspects, WHO, Geneva, Environmental Health Criteria 86, 1989.

45. Lee, X., Benoit, G. and Hu, X., Total gaseous mercury concentration and flux over a coastal saltmarsh vegetation in Connecticut, *Atmos. Environ.*, 34(24), 4205, 2000.

46. Ames, M., Gulen, G. and Olmez, I., Atmospheric mercury in the vapor phase, and in fine and coarse particulate matter at Perch River, New York, *Atmos. Environ.*, 32(5), 865, 1998.

47. Lindberg, S.E. et al., Atmospheric concentrations and deposition of Hg to a deciduous forest at Walker Branch Watershed, Tennessee, *Water, Air, Soil Pollution*, 56, 577, 1991.

48. Grosjean, D., Polycyclic aromatic hydrocarbons in Los Angeles air from samples collected on Teflon, glass, and quartz filters, *Atmos. Environ.*, 17, 2565, 1983.

49. Greenberg, A. et al., Polycyclic aromatic hydrocarbons in New Jersey: A comparison of winter and summer concentrations over a two-year period, *Atmos. Environ.*, 19, 1325, 1985.

50. Menton, R.G., Chuang, J.C. and Chou, Y. L., Comparison study of the 1984, 1987, 1988 winter, and 1988 summer PAH studies, Final Report to U.S. EPA on Contract 68-02-4127, Work Assignment 73; N. K. Wilson, EPA Project Officer; Battelle, Columbus, OH, September 1989.

51. Chuang, J.C., Menton, R.G. and Kuhlman, M.R., Analytical support for the California P-TEAM pre-pilot field study, Final Report to U.S. EPA, Contract 68-02-4127, Work Assignment 74; N. K. Wilson, Project Officer, Battelle, Columbus, OH, September 1989.

52. Chuang, J.C. et al., The U.S. EPA Lake Michigan urban air toxics study: ambient air monitoring and analysis for polycyclic aromatic hydrocarbons, presented at the 1992 EPA/AWMA Symposium on Measurement of Toxic and Related Air Pollutants, Durham, NC, May 4-8, 1992.

TABLE 4.1 (CONTINUED)
Ambient Air Concentrations of 188 Hazardous Air Pollutants

53. Kelly, T.J., Chuang, J.C. and Callahan, P.J., Research for ambient polar volatile organics and semi-volatile phase-distributed organics utilizing TAMS sites, Final Report to U.S. EPA, Contract 68-DO-0007, Work Assignments 1 and 23; J. Pleil, EPA Project Officer, Battelle, Columbus, OH, April 1992.

54. *Asbestiform Fibers: Nonoccupational Health Risks*, published by National Academy Press, National Academy of Sciences, Washington, DC, 1984.

55. Suta B.E. and Levine, R.J., Non-occupational asbestos emissions and exposures, in *Asbestos: Properties, Applications, and Hazards*, Michaels, L. and Chissick, S.S, Eds., John Wiley and Sons, New York, pp 171, 1979.

56. Constant, P. C. Jr., Bergman, F.J. and Atkinson, G.R., Airborne asbestos levels in schools, Final Report to Environmental Protection Agency, prepared by Midwest Research Institute, Contract No. 68-01-5915, 1982.

57. Nicholson, W.J., Measurement of asbestos in ambient air, National Air Pollution Control Administration, Final Report on Contract CPA 70-92, 1971.

58. Wilber, C.G., *Beryllium: A Potential Environmental Contaminant*, published by Charles C. Thomas, Springfield, IL, 1980.

59. Grosjean, D., Liquid chromatography analysis of chloride and nitrate with "negative" ultraviolet detection: ambient levels and relative abundance of gas-phase inorganic and organic acids in southern California, *Environ. Sci. Technol.*, 24:77, 1990.

60. Solomon, P.A. et al., in Preprints of Papers from the National Meeting–American Chemical Society Division of Environmental Chemistry, 28, 72, 1988.

61. Farmer J.C. and Dawson, G.A., Condensation sampling of soluble atmospheric trace gases, *J. Geophys. Res.*, 87, 8931, 1982.

62. Cofer, W.R., Collins, V.G. and Talbot, R.W., Improved aqueous scrubber for collection of soluble atmospheric trace gases, *Environ. Sci. Technol.*, 19, 557, 1985.

63. John, W., Wall, S.M. and Ondo, J.L., A new method for nitric acid and nitrate aerosol measurement using the dichotomous sampler, *Atmos. Environ.*, 22, 1627, 1988.

64. Exposures from the uranium series with emphasis on radon and its daughters, National Council on Radiation Protection and Measurements, NCRP Report No. 77, Bethesda, MD, 1984.

65. Evaluation of occupational and environmental exposures to radon and radon daughters in the United States, National Council on Radiation Protection and Measurements, NCRP Report No. 78, Bethesda, MD, 1984. 66. Pochin, E., *Nuclear Radiation: Risks and Benefits*, Clarendon Press, Oxford, England, 1983.

67. Hawthorne, S.B. et al., PM-10 high-volume collection and quantitation of semi- and nonvolatile phenols, methoxylated phenols, alkanes, and polycyclic aromatic hydrocarbons from winter urban air and their relationship to wood smoke emissions, *Environ. Sci. Technol.*, 26, 2251, 1992.

68. Zankel, K.L. et al., Measurement of ambient ground-level concentrations of hydrogen fluoride, *J. Air Poll. Control. Assoc.*, 37, 1191, 1987.

69. Cooper, H.B. Jr., Rawlings, G.D. and Foote, R.S., Measurement of mercury vapor concentrations in urban atmospheres, *ISA Transactions*, 13:296, 1974.

70. Buchholtz, W.F. and Crow, W.L., Relating SARA Title III emissions to community exposure through ambient air quality measurements, paper 90-170.11, presented at the 83rd Annual Meeting of the A&WMA, Pittsburgh, PA, June 24-29, 1990.

71. Shah, J.J. and Heyerdahl, E.K., National Ambient Volatile Organic Compounds (VOCs) Data Base Update, EPA 600/3-88/010(a), U.S. Environmental Protection Agency, U.S. Government Printing Office, Research Triangle Park, N.C., 1988. (See Shah, J.J. and Singh, H.B., *Environ. Sci. Technol.*, 22, 1381, 1988.)

72. Haggert, B. and Harkov, R., Design and implementation of a module monitoring unit (MMU) to measure ambient volatile organic compounds, paper 84-17.2, presented at the 77th Annual Meeting of the Air Pollution Control Association, San Francisco, CA, June 24-29, 1984.

73. Purdue, L.J. et al., Atlanta ozone precursor monitoring study data report. EPA-600/R-92/157, U.S. EPA, Research Triangle Park, NC, September 1992.

74. McAllister, R. et al., 1990 Urban Air Toxics Monitoring Program. Report No. EPA-450/4-91-024, prepared for U.S. EPA by Radian Corporation, Research Triangle Park, NC, June 1991.

75. Spicer, C.W., Patterns of atmospheric nitrates, sulfate, and hydrogen chloride in the central Ohio River valley over a 1-year period, *Environ. International*, 12, 513, 1986.

76. Spicer, C.W. et al., Atmospheric reaction products from hazardous air pollutant degradation, Report to U.S. EPA, Contract No. 68-02-3169, Work Assignments 33 and 40, by Battelle, Columbus, OH, July 1984.

77. Singh, H.B., Salas, L.J. and Stiles, R.E., Distribution of selected gaseous organic mutagens and suspect carcinogens in ambient air, *Environ. Sci. Technol.*, 16 872, 1982.

78. McAllister, R.A., Epperson, D.L. and Jongleux, R.F., 1989 Urban Air Toxics Monitoring Program: Aldehyde Results, Report No. EPA-450/4-91-006, prepared for U.S. EPA by Radian Corp., Research Triangle Park, NC, January 1991.

79. McAllister, R.A., et al., 1989 Urban Air Toxics Monitoring Program, Report No. EPA-450/4-91-001, prepared for U.S. EPA by Radian Corp., Research Triangle Park, NC, October 1990.

80. Pierotti, D. et al., Measurement of ethylene oxide in ambient air using the Ion Trap Detector™, Finnegan MAT Environmental Analysis Application Data Sheet, Finnegan MAT, San Jose, CA, 1990.

81. Pierotti, D. et al., Measurement of air toxics at environmental background levels, in Proc. 38th ASMS Conf. Mass Spectrometry and Allied Topics, Tucson, AZ, June 1990, published by the American Society for Mass Spectrometry, East Lansing, Michigan, 1990.

82. Lewis, R.G and Lee, R.E. Jr., Air pollution from pesticides: sources, occurrences, and dispersion, in *Air Pollution from Pesticides and Agricultural Processes*, R.E. Lee, Jr., Ed., CRC Press, pp. 5-50, 1976.

83. Fitzgerald, W.F., Vandal, G.V. and Mason, R.P., Atmospheric cycling and air-water exchange of mercury over mid-continental lacustrine regions, *Water, Air, Soil Pollution*. 56, 745, 1991.

84. Smith, R.M. et al., Ambient air and incinerator testing for chlorinated debenzofurans and dioxins by low resolution mass spectrometry, *Chemosphere*, 18, 585, 1989.

85. Smith, R.M. et al., Chlorinated dibenzofurans and dioxins in atmospheric samples from cities in New York, *Environ. Sci. Technol.*, 24(10), 1502, 1990.

86. Ohio Environmental Protection Agency (OEPA), Risk assessment of potential health effects of dioxins and dibenzofurans emitted from Columbus Solid Waste Authority's reduction facility, OEPA Division of Air Pollution Control, February 24, 1994.

87. OEPA, Franklin County Ohio ambient air monitoring study for dioxins and dibenzofurans, OEPA Division of Air Pollution Control, July 27, 1994.

88. OEPA, Dioxin monitoring study 1995—Franklin County Ohio, OEPA Division of Air Pollution Control, September 1995.

89. Connecticut Department of Environmental Protection (CDEP), Ambient monitoring for PCDDs/PCDFs in Connecticut – Fall 1993 through Summer 1994, CDEP final report, Document No. 6350-008-500-R1, September 1995.

90. Hunt, G., Maisel, B. and Zielinska, G., A source of PCDDs/PCDFs in the atmosphere of Phoenix, AZ, *Organohalogen Compounds*, 33, 145, 1997.

91. Dvonch, J.T. et al., An intensive multisite pilot study investigating atmospheric mercury in Broward County, FL, *Water, Air, Soil Pollution*, 80, 169, 1995.

5 Atmospheric Transformation Products of Clean Air Act Title III Hazardous Air Pollutants

5.1 INTRODUCTION

Earlier chapters described the physical and chemical properties of the HAPs, methods that can be used for their measurement, and representative atmospheric concentrations of these species. This chapter deals with the atmospheric reactions that can transform these chemicals into other species, and the processes that remove toxic chemicals from the atmosphere. Figure 5.1 shows schematically some of the many processes that can affect the atmospheric concentration of a toxic air pollutant. These processes can include chemical reactions such as reaction with hydroxyl radical (OH), ozone (O_3), nitrate radical (NO_3), water vapor (H_2O), or other atmospheric constituents. Reactions such as these transform the toxic chemical into a different species. The original toxic molecule is no longer present, but has been replaced by one or more new molecules. Compared with the original HAP, the reaction products of HAP transformations can be more toxic, less toxic, or of similar toxicity.

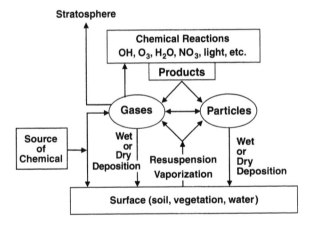

FIGURE 5.1 Chemical and physical processes affecting HAPs in the atmosphere.

Physical processes can also transform or remove toxic pollutants from the air. As noted in Figure 5.1, photolysis (the degradation of a molecule by absorption of light energy), can transform those HAPs that can absorb light in the solar spectrum. The factors affecting photochemical transformations are discussed in detail elsewhere.[1-4] Hazardous air pollutants also can be removed from the atmosphere by deposition on surfaces, including soil, water, and vegetation. HAPs in the particle phase can be removed by direct deposition on surfaces (dry deposition) or by scavenging by cloud or rain droplets (wet deposition). Gas phase HAPs can deposit directly on surfaces, be scavenged by precipitation or clouds, or be adsorbed on particles that are then removed by wet or dry deposition.

These transformation and removal processes affect the fate of the HAP as well as its atmospheric persistence. Human exposure to the HAPs is influenced by the length of time the HAP remains in the atmosphere (persistence) and by its transformation to other chemical products or removal from the air (fate). The Clean Air Act identifies the need for "consideration of atmospheric transformation and other factors which can elevate public health risks from such pollutants." Atmospheric transformations of hazardous pollutants could result in products of higher or lower human health risk. A first step in ascertaining the effect of atmospheric transformations of HAPs on human health involves determining the nature of the transformation products. The reactions and products of some HAPs have been widely investigated, but information on the transformations of many HAPs is scarce or nonexistent. This chapter presents a review and summary of available literature on HAP transformation products. Where possible, we also have summarized information on the atmospheric lifetime of each HAP and the major processes affecting the lifetime. The focus of this chapter is a summary of information on the products of HAP transformations and the persistence of HAPs in the atmosphere. However, a short overview of the experimental methods used to study HAP transformations is presented first to assist the reader in understanding the complexities and uncertainties in the available information for the 188 HAPs.

5.2 EXPERIMENTAL APPROACHES FOR THE STUDY OF HAP TRANSFORMATIONS

The photochemistry and reaction kinetics of atmospheric contaminants have been studied experimentally for several decades, and the investigation of HAP transformations is simply a subset of this broader field. Many of the experimental methods used to estimate reaction rate constants and to determine reaction products for other atmospheric constituents have been applied to the study of the HAPs. We can categorize experimental investigations of HAP transformations into three types:

1. Kinetics studies to determine rate constants
2. Product studies
3. Studies designed to measure both the kinetics and reaction products.

The persistence, or lifetime, of a chemical in the atmosphere can be estimated using the rate constants for the most important atmospheric reactions of the chemical. The exponential lifetime, or natural lifetime, is defined as the time required to reduce the concentration of a chemical to l/e of its original value, where e is the base of natural logarithms (e ≈ 2.718). The rate expression for a typical first order reaction is:

$$\ln\left(\frac{C_o}{C}\right) = kt$$

where C_o is the initial concentration of the chemical, C is the concentration at a later time, k is the reaction rate constant, and t is the reaction time. For the special case where the initial concentration of the chemical is reduced to l/e of its original value,

$$\ln\left(\frac{C_o}{C}\right) = \ln e = 1 = kt$$

and the reaction time t is defined as the lifetime, τ. For this case, the lifetime is simply

$$\tau = 1/k$$

Establishing the true atmospheric lifetime of a chemical would require summing the rate constants for all the pertinent reactions, but frequently one reaction so dominates the removal of a chemical that the chemical's atmospheric lifetime can be approximated using the rate constant for that reaction alone.

For higher order reactions, the concentration of one or more reaction partners is included in the rate expression. For example, many organic chemicals react rapidly with hydroxyl radical in the atmosphere, and the lifetime of the chemical can be approximated by

$$\tau = \frac{1}{k[OH]}$$

However, because [OH] varies widely with the time of day, location, and even season of the year, the assumed concentration of the reaction partner should be noted with the lifetime estimate.

Experimental investigation of reaction kinetics or reaction products typically involves use of a reaction vessel and an analytical system. Depending on the nature of the reaction being investigated and the analytical requirements, the reaction vessel can be large (e.g. a room-size environmental chamber) or small (e.g. a laboratory flask). Larger vessels may be required when the analytical system needs large sample volumes, or when the surface-to-volume ratio must be minimized to reduce the effect of surface reactions. Studies of reaction kinetics that measure the changing concentration of only one or two reactants can often be conducted in small laboratory vessels, whereas investigations that employ a number of measurement approaches to search for reaction products usually need large volumes of sample and require larger reaction vessels. Because much of the progress in the study of HAP transformations has involved use of large environmental chambers, some examples are described below.

Two views of an environmental chamber used to study HAP reactions are shown in Figure 5.2. Sampling and analysis systems on the two sides of the 17 m^3 chamber measure specific reactants or products (e.g., NO, NO_2, CO, formaldehyde, peroxyacetyl nitrate, etc.), while other gas chromatographic and mass spectrometric instruments can search for numerous unknown reaction products or can measure several reactants and products simultaneously. Other instruments measure reaction conditions (temperature, humidity, light intensity, etc.). The large sample volumes required by these instruments dictate the need for a large reaction vessel. Studies with lower sample volume requirements can utilize smaller reaction vessels. Figure 5.3 is an example of a 200-L pyrex chamber with an artificial light source.

The reaction vessels in Figures 5.2 and 5.3 represent fixed volume chambers with artificial light sources to initiate photochemical reactions. Another type of chamber (Figure 5.4) employs large plastic bags of Teflon or similar material. These "variable volume" vessels collapse as sample air is withdrawn, eliminating the need for dilution air required by fixed volume vessels. Also, because the plastic container transmits sunlight well, it can be used outdoors with natural sunlight as the source of radiation for photochemical reactions.

The selection of a reaction vessel for kinetics and reaction product experiments involves many choices and some compromises. Size, shape, surface material, irradiation source, indoor or outdoor use, clean air source, analytical instruments, and cleaning procedures are just a few of the variables that must be considered. Furthermore, a reaction vessel that is well suited to investigate reactions of one class of chemicals may not be ideal for another class. An overview of reaction vessels and experimental methods to study atmospheric reactions can be found in Finlayson-Pitts and Pitts.[4]

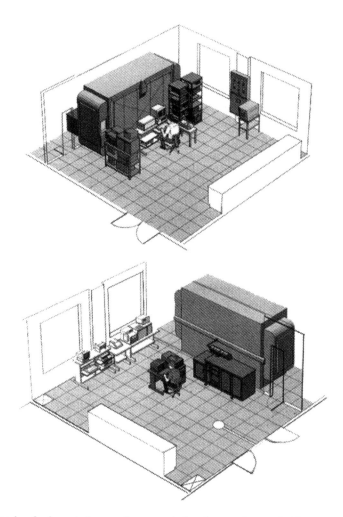

FIGURE 5.2 Example of a large indoor environmental chamber used to study HAP transformation.

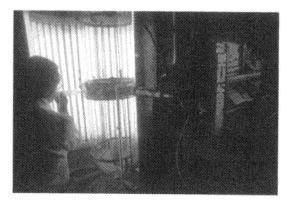

FIGURE 5.3 A small laboratory chamber used to study transformation of HAPs.

FIGURE 5.4 Outdoor Teflon chamber used to study HAP transformations.

5.3 HAZARDOUS AIR POLLUTANT TRANSFORMATIONS

For purposes of this survey, the 188 diverse chemicals designated as HAPs were organized into chemical classes, as described in Chapter 2, to facilitate searching for transformation data. Information on the transformation products of the 188 HAPs was located using a computerized database and through a general review of articles, reference books, proceedings of relevant conferences, and unpublished reports.

Relevant literature was identified through a keyword search of the computerized databases of STN International (Columbus, OH). The databases searched included the Chemical Abstract (CA) files from 1967 to the present, Chemical Abstract Previews (CAP) current files, and the National Technical Information Service (NTIS) files from 1964 to the present.

The search strategy targeted keywords such as "atmospheric or air," "reactions, kinetics or removal," and "rates, constants or lifetime." The search was restricted to English-language citations. The strategy used both abstract and basic index searches to increase the likelihood of finding relevant material.

Using the search strategy described above, master sets of citations were set up in each of the STN files researched. These master sets were then combined with the chemical names and CAS registry numbers to produce citation listings for specific HAP compounds. The listed citations were then reviewed by title and abstract and in their entirety if the initial reviews indicated information of value.

Another resource used to uncover transformation information was the computer database ABIOTIK$_x$.[5] This database was developed to provide the measured reaction rate constants for the degradation of organic compounds in the atmosphere. Upon entering either the compound name or its CAS number, a display is generated identifying published rate constants for several possible atmospheric reactions for that chemical. This database also provides literature citations for the rate data. In this study, the rate data were used to estimate lifetimes and identify significant transformation processes, and the cited literature was reviewed to identify reaction product information.

To assess the impact of atmospheric transformations of HAPs on the risk to human health, it is important to know the products of the transformations, and also whether the rates of the transformations are fast enough to remove the hazardous chemical or to cause the buildup of hazardous levels of the products. Therefore, we searched for information on the lifetimes of the HAPs, and the atmospheric processes that control the lifetimes. For many of the HAPs, the lifetime is likely to be controlled by reaction with hydroxyl radical, while for others, reaction with ozone, photolysis, or wet or dry deposition processes may control or at least influence the lifetime.

Information on the kinetics of atmospheric reactions has enabled us to list the primary removal processes expected to control the lifetimes of many of the HAPs, and to estimate the lifetimes. The lifetimes have been listed in our data summary within three broad ranges:

1. Less than 1 day
2. One to 5 days
3. More than 5 days

These estimates are meant to provide the reader with a sense of the residence times of these species in the atmosphere. The three broad lifetime ranges represent species that are rapidly transformed or removed ($\tau < 1$ day), moderately persistent ($\tau = 1$-5 days), or highly persistent ($\tau > 5$ days). The lifetime, τ, is the exponential lifetime described above, representing the time it takes for the concentration of the HAP to decrease to $1/e$ of its original value.

In some instances, estimates of the atmospheric lifetime of the target chemicals have been reported in the literature. In other cases, we used reported rate constants to identify the most important removal processes and to calculate a corresponding lifetime. For those cases, we assumed the following reactant concentrations in the calculations, to represent long-term average concentrations in a relatively polluted atmosphere:

Reactant	Concentration (molecules/cm³)
O_3	1.5×10^{12}
OH	3.0×10^6
NO_3	2.5×10^9
HO_2	1.0×10^9

Measured rate constants were not available for all of the pertinent reactions. When we could not find an experimentally measured rate constant, we used published rate estimates based on molecular structure. These cases are identified with an asterisk in our data summaries.

In some instances, there was disagreement among rate constants or lifetimes for a selected HAP from multiple references. In these cases, we have listed a range of lifetimes.

Available information on HAP transformation products is compiled in Table 5.1 (see Appendix following Chapter 5). The table is followed by an associated list of 190 citations to relevant literature. The data table lists all 188 HAPs in the same order as in the CAA, giving the name and CAS number of each compound, the chemical formula or structure, the major removal processes, the atmospheric lifetime range, the reported transformation products, references for the reported data, and any additional comments or notes. An asterisk in the last column indicates that the atmospheric lifetime was based on rate constants estimated from structure–reactivity relationships. The transformation products are not listed in any particular order. Many of the references report qualitative information only, so we have not attempted to rank the products by abundance.

A review of Table 5.1 shows that information was found on removal processes, lifetimes, and transformation products for 94 of the 188 HAPs. Twelve compounds were identified as being unlikely to undergo any significant chemical transformation. We were unable to find reported transformation products for 82 compounds that are expected to undergo transformations. We found no relevant information on either atmospheric persistence or products for five compounds. It should be noted that, although transformation products have been reported for 94 of the HAPs, for many of these compounds, the list of products is probably incomplete.

The transformation products listed in Table 5.1 cover a wide range of chemical compositions. Many of the HAP compounds react with OH radical and are degraded to low molecular weight aldehydes, alcohols, organic acids, ketones, nitrates, carbon monoxide, carbon dioxide, and water. Many of the transformation products are multifunctional organic compounds. Some HAPs are transformed to other HAP species, so the degree of health risk depends on the relative toxicity of the original HAP and its transformation products. A further breakdown of HAP transformations is given in Table 5.2, which shows that many of the 82 compounds that may react in the atmosphere, but for which no transformation product data were found, are oxygenated and nitrogenated organic compounds. Even for those chemicals with reported reaction products, the list of products may well be incomplete, depending on the rigor of the experimental study. For example, few studies have reported mass balances to document the completeness of the list of products. This lack of product data represents a serious gap in our knowledge about hazardous air pollutants and affects our ability to assess the health risks posed by their atmospheric transformations. Additional research is needed to elucidate the products and lifetimes of HAPs representing various classes on the HAPs list.

TABLE 5.2
Data Completeness by Compound Class

Compound Class	Data Reported for Removal Process, Lifetime and Transformation Products	No Product Data	No Transformation Anticipated	No Data
Hydrocarbons	3	0	0	0
Halogenated Hydrocarbons	19	7	1	0
Aromatic Compounds	16	3	0	1
Halogenated Aromatics	5	3	0	0
Nitrogenated Organics	14	36	0	3
Oxygenated Organics	23	15	0	0
Pesticides/Herbicides	4	11	0	1
Inorganics	18	4	11	0
Phthalates	0	3	0	0
Sulfates	2	0	0	0
Total	94	82	12	5

The process that drives the transformation of many of the HAP compounds in the atmosphere is reaction with hydroxyl radical (OH). Of the 181 HAPs for which a tropospheric removal process is reported in Table 5.1, 86% show reaction with hydroxyl as an important removal mechanism. This is consistent with the hydroxyl radical's known role as an "atmospheric cleanser." Thompson[6] has reviewed the hydroxyl radical and other species that control the atmosphere's oxidizing capacity.

Reactions with ozone and nitrate radical contribute to the removal of a number of the HAPs, but these reactions appear to control the lifetimes of only a few of the species. Photolysis is also an important degradation mechanism for a number of the HAPs, and is thought to be the primary removal mechanism for 17 chemicals (one of these, CCl_4, is removed by photolysis in the stratosphere). Deposition was identified as the primary removal process for a number of the HAPs, especially those that exist in particle form and those that are not expected to undergo significant chemical transformation in the atmosphere. Consequently, their removal is expected to be controlled by physical removal processes (wet and dry deposition). For seven of the HAPs, reaction with or in liquid water is listed as the major removal pathway. The lifetimes of these species are expected to be controlled by this reaction when liquid water is present (in the form of clouds, precipitation, over bodies of surface water) but to be dominated by other removal mechanisms in the absence of liquid water. One of the HAPs (radionuclides including radon) is removed by radioactive decay or deposition.

The persistence of a hazardous chemical in the atmosphere influences the route by which humans may be exposed to it and also the extent of population exposure. A chemical with a very short lifetime will often be transformed to products or removed from the air before large populations are exposed, whereas very persistent chemicals may be transported over large distances (sometimes even globally) with the concomitant exposure of large numbers of people via inhalation. Of course, the location and distribution of emission sources also plays a major role in population exposure, but the atmospheric lifetime is an important factor.

We classified the atmospheric lifetimes of the HAPs into the three previously described broad ranges in Table 5.1. Seventy-one of the HAPs (38%) are reported to have lifetimes of less than 1 day. Another 27 compounds (14%) have intermediate lifetimes of 1 to 5 days, while 56 (30%) are fairly persistent, with lifetimes greater than 5 days. There are 21 HAPs for which lifetime estimates are inconsistent. In these cases, the range of lifetime estimates is listed in the table. For eight HAPs, the estimated atmospheric lifetime depends on the phase the chemical is in. For many semivolatile species, the fraction existing in the gas phase is removed more rapidly than that found in the particle phase.

No lifetime estimates were found for five chemicals. These species are: acrylamide, chloramben, dimethylformamide, hexamethylphosphoramide, and coke oven emissions. The latter HAP has no lifetime estimate due to the unspecified nature of the chemicals emitted.

The lifetime estimates suggest that hydrocarbons, nitrogenated organics, aromatic compounds, sulfates, and pesticides and herbicides are generally expected to degrade fairly rapidly in the atmosphere. The oxygenated organics are distributed evenly across the lifetime ranges. The inorganics, halogenated hydrocarbons, and halogenated aromatics are reported to be relatively persistent in the atmosphere. As a cautionary note, we find in Table 5.1 that comparisons of lifetimes among HAPs with structural similarities sometimes show inconsistencies. The lifetimes reported in this document have been measured or estimated by a variety of methods, which have not been reviewed for consistency. Consequently, the lifetime estimates provided here should be used with caution.

5.4 TRANSFORMATIONS OF 33 HIGH PRIORITY HAPS

As described in Chapter 1, EPA has targeted 33 of the 188 HAPs as presenting the greatest threat to public health in urban areas.[7] These 33 HAPs, along with diesel particulate matter, are being included in a national assessment of toxic air pollutants to identify those that present the greatest risk to public health.

An important question is how atmospheric reactions of the 33 high priority HAPs affect the public health risk associated with these chemicals. Clearly, a knowledge of the lifetimes of these species is important in assessing population exposure and the risk posed by these chemicals. But it is also important to include the toxicity of their transformation products in any risk assessment. Atmospheric reactions can transform some HAPs into innocuous products, reducing the risks posed by these chemicals. But transformations can also produce pollutants more toxic than the original HAP, and this transformation needs to be included in the risk assessment process.

An overview of our knowledge about the atmospheric transformations of the 33 high priority HAPs is provided in Table 5.3. The compounds on this list cover a broad range of chemical categories, lifetimes, and transformation products. Twenty-four of these compounds are expected to be fairly persistent, with lifetimes of a day or longer. Reaction with hydroxyl radical is expected to be a major removal pathway for most of the 33 HAPs, but several other transformation and removal routes are cited.

The last column in Table 5.3 indicates whether specific transformation products have been identified for each compound. We found no information about atmospheric transformation products for propylene dichloride, hexachlorobenzene, quinoline, or 1,1,2,2-tetrachloroethane. This represents an important gap in our knowledge about these high priority HAPs, and will add uncertainty to the national assessment of the risks posed by these chemicals.

TABLE 5.3
Overview of Lifetime and Transformation Data for 33 High Priority HAPs

Compound	CAS Registry No.	Major Removal Process	Atmospheric Lifetime (days)	Transformation Products
Acetaldehyde	75-07-0	OH	<1	Yes
Acrolein	107-02-8	OH	<1	Yes
Acrylonitrile	107-13-1	OH	1-5 to .5	Yes
Arsenic compounds	--	Deposition	>5	No transformation
Benzene	71-43-2	OH	>5	Yes
Beryllium compounds	--	Deposition	>5	No transformation
1,3-Butadiene	106-99-0	OH, O_3	<1	Yes
Cadmium compounds	--	Deposition	>5	No transformation
Carbon tetrachloride	56-23-5	Stratospheric photolysis	>5	No transformation (in troposphere)
Chloroform	67-66-3	OH	>5	Yes
Chromium compounds	--	Deposition	>5	No transformation
Coke oven emissions	--	*	(a)	(a)
1,3-Dichloropropene	542-75-6	OH, O_3	<1	Yes
Ethylene dibromide	106-93-4	OH	>5	Yes
Ethylene dichloride	107-06-2	OH	>5	Yes
Ethylene oxide	75-21-8	OH	>5	Yes
Formaldehyde	50-00-0	Photolysis	<1	Yes
Hexachlorobenzene	118-74-1	OH	>5	No information
Hydrazine	302-01-2	OH	<1	Yes
Lead compounds	--	Deposition (inorganic compounds)	>5	No transformation
		OH, O_3 (organo-lead compounds)	<1 to 1-5	Yes
Manganese compounds	--	Deposition	>5	No transformation
Mercury compounds	--	Aqueous OH (elemental) Deposition (particle phase)	>5	Yes
Methylene chloride	75-09-2	OH	>5	Yes
Nickel compounds	--	Deposition	>5	No transformation
Polychlorinated biphenyls	1336-36-3	Photolysis, OH	>5	Yes
Polycyclic organic matter**	e.g. 85-01-8	OH	<1 to 1-5	Yes
Propylene dichloride	78-87-5	OH	>5	No information
Quinoline	91-22-5	OH	1 to 5	No information
2,3,7,8-Tetrachlorodibenzo-p-dioxin	1746-01-06	Photolysis, deposition	1-5 to >5 (gas phase) >5 (particle phase)	Yes
1,1,2,2-Tetrachloroethane	79-32-5	OH	>5	No information
Tetrachloroethylene	127-18-4	OH	>5	Yes
Trichloroethylene	79-01-6	OH, O_3	<1 to 1-5	Yes
Vinyl chloride	75-01-4	OH, O_3	<1 to 1-5	Yes

* Specific chemicals are not identified, so transformation information cannot be cited.
**Information cited for the representative compound phenanthrene.

5.5 TRANSFORMATIONS OF OTHER ATMOSPHERIC CHEMICALS

Besides transformations of the 188 HAPs, other atmospheric transformations can elevate the risk to public health. Many chemicals that are emitted to the atmosphere, whether from anthropogenic or natural sources, are not on the Title III list of 188 chemicals, and may not be toxic themselves, but can undergo atmospheric transformations that generate toxic products, some of which might be HAPs. Although examination of non-HAP transformations is outside the scope of this chapter, such transformations should be considered in assessing human exposure to toxic air pollutants. In the paragraph below, one type of study addressing this issue is reviewed. This type of research is investigating mutagenic activity from the atmospheric transformations of non-HAP compounds. The significance of the findings with respect to public health risk merits our attention.

As noted above, there are non-HAP and even nontoxic compounds that undergo atmospheric transformations to generate chemicals that may present human health hazards. An example of such a compound is the simple hydrocarbon propene. It is reported that this ubiquitous ambient air constituent, when irradiated in the presence of NO_x, yields transformation products that are mutagens. The products that have been identified from the atmospheric transformation of propene include formaldehyde, acetaldehyde, peroxyacetyl nitrate, nitric acid, propylene glycol dinitrate, 2-hydroxy propyl nitrate, 2-nitropropyl alcohol, α-nitroacetone, and carbon monoxide.[8] These products do not account for all of the mutagenic activity; other unidentified mutagens likely include organic peroxides and nitrates. Further investigations of the transformation of propene by reaction with O_3,[9] and with hydroxyl and nitrate radicals[10] also identify organic oxygenates as products, although they do not account for the mutagenic activity associated with propene/NO_x transformations. This work demonstrates that compounds considered to be nontoxic can undergo atmospheric transformations to produce HAPs, as well as other toxic pollutants. These studies were able to demonstrate mutagenic activity, even though the specific mutagens could not always be identified. This approach warrants careful consideration as a means of investigating the risks associated with atmospheric transformations.

5.6 SUMMARY

This chapter has summarized available literature on the transformation products and atmospheric persistence of the 188 hazardous air pollutants listed in the Clean Air Act. The atmospheric lifetimes for all but five of the chemicals have been estimated, although there is inconsistency in the estimates for 21 of the HAPs. The transformation product information is much less complete. Transformation products have been identified for only half of the 188 chemicals.

The transformation products formed during the atmospheric reactions of HAPs include aldehydes, alcohols, peroxides, nitrosamines, nitramines, amides, organic acids, ketones, nitrates, carbon monoxide, carbon dioxide, and a wide variety of other oxygenated, nitrogenated, halogenated, or sulfur-containing species. Many of the products are multifunctional chemicals. Some of the products are known to be toxic, while many others have never been tested for toxicity. Reaction products have not been identified for 38 of the chemicals that are expected to react rapidly in the atmosphere (lifetime < 1 day). Because these chemicals may be transformed to potentially toxic products before atmospheric dilution has reduced their concentrations to negligible levels, knowledge of the atmospheric reaction products is a priority.

Many of the 188 HAPs undergo transformations to produce other hazardous chemicals on the HAPs list. For example, formaldehyde is a product of many HAP transformations. Transformation products can be either more or less toxic than the original HAP. For example, one product of chloroform oxidation (phosgene) is reported to be significantly more toxic than chloroform itself.[11] In addition, chemicals not currently identified as hazardous air pollutants can undergo atmospheric transformations to generate toxic products (including HAPs) that pose a potential health risk to humans.

Considerably more information must be gathered on the atmospheric reactions of many of the HAPs before their transformations can be fully incorporated in health risk assessments. As a starting point, efforts should be focused on the five HAPs for which neither lifetime nor product data are available, and on those 38 HAPs that may undergo rapid atmospheric transformations, but for which no transformation product data are available. The transformation products of this latter group of chemicals may build up relatively rapidly, and therefore could represent a health risk if they are toxic. On the other hand, if the products are found to be nontoxic, then the rapid removal of these HAPs from the atmosphere should also be taken into account in the risk assessment process. The specific chemicals in these two groups are listed in Table 5.4. The elucidation of the transformation products of these HAPs should be given a high priority.

TABLE 5.4
Recommended Priority HAPs for Transformation Product Studies

Rapid Transformation/No Reaction Product Data

Acetamide	3,3′-Dimethoxybenzidine
2-Acetylaminofluorene	Dimethylaminoazobenzene
Acrylic acid	3,3′-Dimethylbenzidine
4-Aminobiphenyl	Dimethylcarbamoyl chloride
o-Anisidine	1,2-Diphenylhydrazine
Benzidine	Ethyl carbamate
Bis(2-ethylhexyl) phthalate (DEHP)	Ethyleneimine
Calcium cyanamide	Heptachlor
Captan	Isophorone
Carbaryl	Lindane
Catechol	Methoxychlor
Chlordane	Methyl isocyanate
Chlorobenzilate	4,4′-Methylene bis (2-chloroaniline)
DDE	4,4′-Methylenedianiline
Diazomethane	p-Phenylenediamine
Dibutyl phthalate	Propoxur
3,3'-Dichlorobenzidine	1,2-Propyleneimine
Dichlorvos	2,4-Touenediamine
Diethanolamine	o-Toluidine

No Lifetime or Reaction Product Data

Acrylamide	Hexamethylphosphoramide
Chloramben	Coke oven emissions
N,N-Dimethylformamide	

REFERENCES

1. Leighton, P.A., *Photochemistry of Air Pollution*, Academic Press, New York, 1961.
2. Calvert, J.G. and Pitts, J.N. Jr., *Photochemistry*, Wiley, New York, 1966.
3. Heicklen, J., *Atmospheric Chemistry*, Academic Press, New York, 1976.
4. Finlayson-Pitts, B.J. and Pitts, J.N. Jr., *Chemistry of the Upper and Lower Atmosphere Theory, Experiments, and Applications*, Academic Press, San Diego, 2000.
5. Daniel, B., Merz, G. and Klöpffer, W., ABIOTIK$_x$ Data-Base, Daniel Electronic, in behalf of Battelle Institut e.V., Frankfurt, Germany, November 1990.
6. Thompson, A.M., The oxidizing capacity of the earth's atmosphere: probable past and future changes, *Science*, 256, 1157, 1992.
7. U.S. EPA, National Air Toxics Program: The Integrated Urban Strategy, 64 FR 38705, July 19, 1999. Available at: http://www.epa.gov/ttnatw01/urban/urbanpg.html.
8. Kleindienst, T.E. et al., The mutagenic activity of the products of propylene photooxidation, *Environ. Sci. Technol.*, 19, 620, 1985.

9. Shepson, P.B. et al., The production of organic nitrates from hydroxyl and nitrate radical reaction with propylene, *Environ. Sci. Technol.*, 19, 849, 1985.

10. Shepson, P.B. et al., The mutagenic activity of the products of ozone reaction with propylene in the present and absence of nitrogen dioxide, *Environ. Sci. Technol.*, 19, 1094, 1985.

11. Sittig, M., *Handbook of Toxic and Hazardous Chemicals and Carcinogens*, 2nd ed., Noyes Publications, Park Ridge, NJ, 1985.

APPENDIX

TABLE 5.1
Summary of Hazardous Air Pollutant Transformations and Lifetimes (Chemicals shown in italics are high priority urban HAPs)

Compound and CAS Number	Chemical Formula/Structure	Major Removal Processes	Reported Atmospheric Lifetimes (days)	Transformation Products	References	Comments/Notes
Acetaldehyde *75-07-0*	CH_3CHO	*OH, photolysis*	*<1*	*Formaldehyde, carbon dioxide, carbon monoxide, PAN, hydrogen peroxide, methyl nitrate*	*1,2,4,5,13,14,16, 17,19,20,29,52, 54,55,69,79*	*
Acetamide 60-35-5	$CH_3C(O)NH_2$	OH	<1	No information found	21	
Acetonitrile 75-05-8	CH_3CN	OH	>5	No information found	19,21,52,54,55, 69,101,102,106	
Acetophenone 98-86-2		OH	1 to 5	No information found	55	
2-Acetylaminofluorene 53-96-3		OH	<1	No information found	21	*

TABLE 5.1
Summary of Hazardous Air Pollutant Transformations and Lifetimes (Chemicals shown in italics are high priority urban HAPs)

Compound and CAS Number	Chemical Formula/Structure	Major Removal Processes	Reported Atmospheric Lifetimes (days)	Transformation Products	References	Comments/Notes
Acrolein 107-02-8	*H₂C=CHCHO*	*OH, photolysis, O₃*	*<1*	*Formaldehyde, carbon dioxide, glyoxal, formic acid, peroxyacylnitrate*	*2,3,5,9,13,21,50, 52,54,55,79,93, 108,145,146, 166*	Reaction with OH and O₃ likely
Acrylamide 79-06-1	CH₂CHC(O)NH₂	No information found	No information found	No information found		
Acrylic Acid 79-10-7	H₂C=CHCO₂H	OH, deposition	<1	No information found	17,21	
Acrylonitrile 107-13-1	*CH₂CHCN*	*OH*	*1-5 to >5*	*Formaldehyde, formic acid, hydrogen cyanide, formyl cyanide, oxalic acid mononitrile*	*5,7,11,13,19,21, 52,102,103, 108,147*	
Allyl chloride (3-chloro-1-propene) 107-05-1	CH₂=CHCH₂Cl	OH, O₃	<1	Formaldehyde, formic acid, chloroacetaldehyde, acetaldehyde, 1,3-dichloroacetone, acrolein, glyoxal, PAN, Cl-PAN, chlorinated hydroxy carbonyls, formyl chloride	5,21,13,54,55, 93,137,146, 148,149,150	
4-Aminobiphenyl 92-67-1	(structure: biphenyl with NH₂)	OH	<1	No information found	21	*

Compound	Structure	Mechanism		Transformation Products	Ref.	
o-Anisidine 90-04-0	NH_2, OCH_3	OH	<1	No information found	21	*
Aniline 62-53-3	NH_2	OH	<1	Nitrosamines, nitramines, nitrobenzene, formic acid, hydrogen peroxide, nitrophenols, nitrosobenzene, benzidine, amino-phenol, nitroaniline, phenol, hydroxybenzonitrile, N-phenyl-formamide	7,8,9,55,102	
Asbestos 1332-21-4	Silicate minerals of the serpentine and amphibole groups	Deposition	>5	No chemical transformation	112	
Benzene 71-43-2	(benzene ring)	OH, NO_3	>5	Phenol, nitrobenzene, nitrophenol, dinitrophenol, glyoxal, 2-butene-1,4-dial	7,8,9,14,17,19, 21,35,41,52,54, 55,65,151,189	*
Benzidine 92-87-5	H_2N—(biphenyl)—NH_2	OH	<1	No information found	21	*
Benzotrichloride 98-07-7	Cl—C—Cl with Cl (phenyl)	OH	>5	No information found	21	

TABLE 5.1 (CONTINUED)
Summary of Hazardous Air Pollutant Transformations and Lifetimes (Chemicals shown in italics are high priority urban HAPs)

Compound and CAS Number	Chemical Formula/Structure	Major Removal Processes	Reported Atmospheric Lifetimes (days)	Transformation Products	References	Comments/Notes
Benzyl chloride 100-44-7	$Cl-CH_2$ (benzyl ring)	OH	1 to >5	Phenol, chloromethyl phenols, ring cleavage products, benzaldehyde, nitrobenzyl chloride, chloromethyl glyoxal, peroxybenzoyl nitrate	5,21,31,52,54, 55,103,146, 148,152	
Biphenyl 92-52-4	(biphenyl structure)	OH	1 to 5	3-Nitrobiphenyl, 2,3,4-hydroxybiphenyl, 2-hydroxybiphenyl	19,21,23,45,46, 47,48,49,50,52, 55,78,166	
Bis(2-ethylhexyl) phthalate (DEHP) 117-81-7	C_2H_5, C_5H_{11} $O=C-OCH_2-C_5H_{11}$ C_2H_5 (phthalate ester structure)	OH, deposition	<1	No information found	21	*
Bis(chloromethyl) ether 542-88-1	$ClCH_2OCH_2Cl$	Liquid H_2O, OH	<1 to 1-5	Hydrochloric acid, formaldehyde, chloromethylformate	5,21,130,131, 167	Rapid reaction with liquid H_2O

Compound	Structure	Process	Persistence (days)	Products	References	Notes
Bromoform 75-25-2	CHBr$_3$	OH, photolysis	>5	No information found	166	See chloroform for analogous pathways
1,3-Butadiene 106-99-0	H$_2$C=CHCH=CH$_2$	OH, O$_3$	<1	Acrolein, formaldehyde, glyoxal, trans-4-nitroxy-2-butenal, 2-butanone, 1-nitroxy-3-buten-2-one, cyclohexanone, glycolaldehyde, furan, 1,3-butadiene monoxide, glycidaldehyde, 1,3-butadiene diepoxide, 3-hydroxy-propanaldehyde, propanal, hydroxyacetone, malonaldehyde	15,17,21,50,52, 54,55,108,138, 145,153,162, 163,165	
Calcium cyanamide 156-62-7	CaNCN	OH, deposition	<1 (gas phase) >5 (particle phase)	No information found	21,166	*
Captan 133-06-2		OH	<1	No information found	21	*
Carbaryl (Sevin) 63-25-2		OH, deposition	<1 (gas phase) >5 (particle phase)	No information found	21,166	*
Carbon disulfide 75-15-0	CS$_2$	OH	1-5 to >5	Carbonyl sulfide, sulfur dioxide	17,19,26,119	Considerable disagreement in reported OH rate constants
Carbon tetrachloride 56-23-5	CCl$_4$	Photolysis in stratosphere	>5	No tropospheric reactions anticipated	5,7,15,21,55,69	

TABLE 5.1 (CONTINUED)
Summary of Hazardous Air Pollutant Transformations and Lifetimes (Chemicals shown in italics are high priority urban HAPs)

Compound and CAS Number	Chemical Formula/Structure	Major Removal Processes	Reported Atmospheric Lifetimes (days)	Transformation Products	References	Comments/Notes
Carbonyl sulfide 463-58-1	COS	OH	>5	Carbon dioxide, hydrogen sulfide, sulfur dioxide	19,26,27,69	
Catechol 120-80-9		OH	<1	No information found	21	*
Chloramben 133-90-4		No information found	No information found	No information found		
Chlordane 57-74-9		OH	<1 to >5	No information found	21,136,166	Considerable disagreement over lifetime
Chlorine 7782-50-5	Cl_2	Photolysis	<1	Hydrogen chloride	111	
Chloroacetic acid 79-11-8	$ClCH_2CO_2H$	OH, deposition	>5	No information found	21	*

Compound	Structure	Reactant	Lifetime	Products	References	Notes
2-Chloroacetophenone 532-27-4	(benzene ring with C(=O)–CH$_2$Cl)	OH	>5	No information found	21	*
Chlorobenzene 108-90-7	(benzene ring with Cl)	OH	>5	Chlorophenols, 1-chloro-3-nitrobenzene, ring cleavage products, chloronitrophenols	5,7,8,9,17,19,21, 23,42,50,55, 146,152	
Chlorobenzilate 510-15-6	(structure with two Cl-phenyl groups, OH, C(=O)–O–CH$_3$)	OH	<1 to >5	No information found	21,166	* Considerable disagreement over lifetime
Chloroform 67-66-3	CHCl$_3$	OH	>5	Phosgene, hydrogen chloride	5,7,14,15,17,21, 22,54,55,69, 129	
Chloromethyl methyl ether 107-30-2	ClCH$_2$OCH$_3$	Liquid H$_2$O, OH	<1 to 1-5	Chloromethyl formate, methyl formate, methanol, chloromethanol	5,21,130,167	Rapid reaction with liquid H$_2$O
Chloroprene (2-chloro 1,3-butadiene) 126-99-8	CH$_2$=CHCCl=CH$_2$	OH, O$_3$	<1	Formaldehyde, ethanediol, chlorohydroxy acids, aldehydes, formic acid	5,21,146,149	
Cresol/Cresylic Acid 1319-77-3	(benzene ring) + CH$_3$ + OH	OH, NO$_3$	<1	Nitrocresols, ring cleavage products	5,21	

TABLE 5.1 (CONTINUED)
Summary of Hazardous Air Pollutant Transformations and Lifetimes (Chemicals shown in italics are high priority urban HAPs)

Compound and CAS Number	Chemical Formula/Structure	Major Removal Processes	Reported Atmospheric Lifetimes (days)	Transformation Products	References	Comments/Notes
o-Cresol 95-48-7		OH, NO_3	<1	Nitrocresol, 2-hydroxy-benzaldehyde, dihydroxy-toluenes,	4,7,17,21,43,52, 54,55,60,102, 103,117	
m-Cresol 108-39-4		OH, NO_3	<1	3-Methyl-2-nitrophenol, 5-methyl-2-nitrophenol, hydroxynitrotoluene, dihydroxytoluene	17,21,52,54,103, 117	
p-Cresol 106-44-5		OH, NO_3	<1	4-Methyl-2-nitrophenol, hydroxynitrotoluenes, dihydroxytoluenes	8,17,21,52,54, 55,60,103,117	
Cumene (isopropyl benzene) 98-82-8		OH	1-5	Isopropylphenols	21,55,168	See toluene for analogous products

Compound	Structure	Process	%	Product	References	
2,4-D (including salts and esters) (e.g., 2,4-Dichlorophenoxy-acetic acid 94-75-7)	(structure: 2,4-dichlorophenoxyacetic acid, O–CH$_2$–COOH)	OH	<1	2,4-dichloroanisole	21,33	*
DDE (1,1-Dichloro-2,2-bis(p-chlorophenyl)ethylene) 72-55-9	(structure: CH=CCl$_2$ with two chlorophenyl groups)	Photolysis, OH	<1	No information found	21	
Diazomethane 334-88-3	CH$_2$N$_2$	Photolysis, O$_3$	<1	No information found	52	*
Dibenzofuran 132-64-9	(structure: dibenzofuran)	OH	1-5	Nitrodibenzofuran	21,30,31,169	*
1,2-Dibromo-3-chloropropane 96-12-8	BrCH$_2$BrCHCH$_2$Cl	OH	>5	1,2-Dibromopropanol, chlorobromopropanol	21,32,50	
Dibutylphthalate 84-74-2	(structure: dibutyl phthalate, two C–OC$_4$H$_9$ esters)	OH, deposition	<1	No information found	21	*

TABLE 5.1 (CONTINUED)
Summary of Hazardous Air Pollutant Transformations and Lifetimes (Chemicals shown in italics are high priority urban HAPs)

Compound and CAS Number	Chemical Formula/Structure	Major Removal Processes	Reported Atmospheric Lifetimes (days)	Transformation Products	References	Comments/Notes
1,4-Dichlorobenzene (p-dichlorobenzene) 106-46-7		OH	>5	2,5-Dichloro-6-nitrophenol, 2,5-dichloronitrobenzene, 2,5-dichlorophenol, chlorinated phenols, nitro compounds, ring cleavage products	5,7,17,19,21,51, 54,55,56,70,72	*
3,3′-Dichlorobenzidine 91-94-1		Deposition, Photolysis, OH	<1 (gas phase) >5 (particle phase)	No information found	21,166	
Dichloroethyl ether (Bis[2-chloroethyl] ether) 111-44-4	$(ClCH_2CH_2)_2O$	OH	1 to 5	No information found	21	*
1,3-Dichloropropene 542-75-6	*$CH_2ClCH=CHCl$*	*OH, O_3*	*<1*	*Formylchloride, chloroacetalde-hyde, chloroacetic acid, carbon dioxide, carbon monoxide, hydrogen chloride*	*17,21,34,55,74, 75,170*	
Dichlorvos 62-73-7		OH	<1	No information found	95	*

Compound / CAS	Structure	Process	Lifetime	Products	References	Notes
Diethanolamine 111-42-2	$(HOC_2H_4)_2NH$	OH	<1	No information found	17,21	*
Diethyl sulfate 64-67-5	$(C_2H_5)_2SO_4$	Liquid H_2O, OH	<1	Monoethyl sulfate, sulfuric acid, ethanol	100,170	Products based on hydrolysis
3,3'-Dimethoxybenzidine 119-90-4	H_2N–(ring, OCH_3, CH_3O)–NH_2	OH, photolysis	<1	No information found	21	*
4-Dimethylaminoazobenzene 60-11-7	$(CH_3)_2N$–C6H4–N=N–C6H5	Deposition, photolysis, OH	<1	No information found	21	*
N,N-Dimethylaniline 121-69-7	CH_3–N(CH_3)–C6H5	OH	<1	Nitrosamines, nitramines, N-methyl formanilide, formic acid, formaldehyde, hydrogen peroxide	9,55,102,128	
3,3'-Dimethylbenzidine 119-93-7	H_2N–(ring, CH_3, H_3C)–NH_2	OH	<1	No information found	21	*
Dimethylcarbamoyl chloride 79-44-7	$(CH_3)_2NC(O)Cl$	OH	<1	No information found	21	*
N,N-Dimethylformamide 68-12-2	$HC(O)N(CH_3)_2$	No information found	No information found	No information found		

TABLE 5.1 (CONTINUED)
Summary of Hazardous Air Pollutant Transformations and Lifetimes (Chemicals shown in italics are high priority urban HAPs)

Compound and CAS Number	Chemical Formula/Structure	Major Removal Processes	Reported Atmospheric Lifetimes (days)	Transformation Products	References	Comments/Notes
1,1-Dimethylhydrazine 57-14-7	$(CH_3)_2NNH_2$	O_3,OH, NO_2, NO	<1	Dimethyl nitrosamine, HONO, N-nitrosohydrazine, tetramethyl tetrazene	21,109	*
Dimethyl phthalate 131-11-3		OH, deposition	1 to 5	No information found	21	*
Dimethyl sulfate 77-78-1	$(CH_3)_2SO_4$	Liquid H_2O; OH	<1	Monomethyl hydrogen sulfate, sulfuric acid, methanol	99	Products based on hydrolysis
4,6-Dinitro-o-cresol, and salts (e.g., 534-52-1)		Photolysis deposition	>5	No information found	21,166	*

Compound	Structure	Process	Lifetime	Products	References	
2,4-Dinitrophenol 51-28-5	(structure)	Photolysis, deposition	>5	No information found	21,166	*
2,4-Dinitrotoluene 121-14-2	(structure)	Photolysis	1 to 5	No information found	21	*
1,4-Dioxane 123-91-1	(structure)	OH	1 to 5	Formaldehyde, 2-oxydioxane	21,28,55,146, 154	
1,2-Diphenylhydrazine 122-66-7	(structure)	OH	<1	No information found	21	*
Epichlorohydrin 106-89-8	(structure)	OH	>5	Formaldehyde, glyoxylic acid, PAN (possibly), chloroacetalde-hyde, peroxychloroacetyl nitrate	5,7,17,21,40,55	
1,2-Epoxybutane 106-88-7	(structure)	OH	1 to 5	No information found	8,55,73,86	

TABLE 5.1 (CONTINUED)
Summary of Hazardous Air Pollutant Transformations and Lifetimes (Chemicals shown in italics are high priority urban HAPs)

Compound and CAS Number	Chemical Formula/Structure	Major Removal Processes	Reported Atmospheric Lifetimes (days)	Transformation Products	References	Comments/Notes
Ethyl acrylate 140-88-5	$H_2C=CHCO_2C_2H_5$	O_3, OH	1 to 5	Glyoxylic acid, ethyl glyoxylate, ethyl formate, epoxides	11,21	
Ethylbenzene 100-41-4	C_2H_5 (benzene ring)	OH	1 to 5	Glyoxal, ethylglyoxal, acetophenone, formaldehyde, acetaldehyde, peroxyacetyl nitrate, peroxypropionyl nitrate, butenedial, 4-oxo-2-hexenal, 2-ethyl-butenedial, 4-oxo-2-butenoic acid, 4-oxo-2-hexenoic acid, 3-methyl-2,5-furandione 2,5-furandione	17,21,31,54,55, 110,166,179	
Ethyl carbamate 51-79-6	$NH_2C(O)OC_2H_5$	OH	<1	No information found	21	*
Ethyl chloride (Chloroethane) 75-00-3	CH_3CH_2Cl	OH	>5	No information found	14,17,21,55	
Ethylene dibromide (1,2-Dibromoethane) 106-93-4	*CH_2BrCH_2Br*	*OH*	*>5*	*Formaldehyde, bromoethanol, hydrogen bromide, formyl bromide*	*5,14,17,21,55,76*	

Compound	Structure	Process	Lifetime	Products	References
Ethylene dichloride (1,2-dichloroethane) 107-06-2	CH_2ClCH_2Cl	*OH*	>5	*Formyl chloride, chloroacetyl chloride, hydrogen chloride, chloroethanol*	7,14,15,17,21, 55,76
Ethylene glycol 107-21-1	$HOCH_2CH_2OH$	OH, deposition	1 to 5	Hydroxyacetaldehyde	17,21,55,166
Ethyleneimine 151-56-4	(structure)	OH	<1 to 1-5	No information found	21,55
Ethylene oxide 75-21-8	(structure)	*OH*	>5	*Methanol, nitromethane, methyl nitrate, ethyl nitrate*	4,5,7,9,15,17,19, 21,55
Ethylene thiourea 96-45-7	(structure)	Deposition	>5	No information found	166
Ethylidene dichloride (1,1-Dichloroethane) 75-34-3	CH_3CHCl_2	OH	>5	Acetyl chloride	14,21,39,55,76, 166
Formaldehyde 50-00-0	*HCHO*	*Photolysis, OH*	<1	*Carbon monoxide, carbon dioxide, hydrogen peroxide, hydrogen*	4,5,9,13,16,17, 20,21,52,53,54, 55,69,79,91
Heptachlor 76-44-8	(structure)	OH, photolysis	<1	No information found	21

TABLE 5.1 (CONTINUED)
Summary of Hazardous Air Pollutant Transformations and Lifetimes (Chemicals shown in italics are high priority urban HAPs)

Compound and CAS Number	Chemical Formula/Structure	Major Removal Processes	Reported Atmospheric Lifetimes (days)	Transformation Products	References	Comments/Notes
Hexachlorobenzene 118-74-1		*OH*	*>5*	*No information found*	*21*	*
Hexachlorobutadiene 87-68-3	$Cl_2C=CCl-CCl=CCl_2$	OH, NO_3, O_3	>5	No information found	21,166	*
1,2,3,4,5,6- Hexachlorocyclohexane and isomers (e.g., Lindane 58-89-9)		OH, photolysis	<1 to >5	Hexachlorocyclohexane (photoisomerization)	21,33,166	* Significant disagreement over lifetime
Hexachlorocyclopentadiene 77-47-4		OH, O_3	<1	Phosgene, diacylchlorides, ketones, hydrogen chloride	5,21,149	*
Hexachloroethane 67-72-1	Cl_3CCCl_3	No tropospheric removal	>5	No information found	21,166	*
Hexamethylene-diisocyanate 822-06-0	$OCN(CH_2)_6NCO$	OH	1 to 5	No information found	166	

Compound / CAS	Structure	Loss process	Lifetime	Products	References	Notes
Hexamethyl-phosphoramide 680-31-9	$[(CH_3)_2N]_3PO$	No information found	No information found	No information found		
Hexane 110-54-3	$CH_3(CH_2)_4CH_3$	OH, NO_3	<1	2-Hexylnitrate, 3-hexylnitrate, aldehydes, ketones	35,36,54,55,92, 102	
Hydrazine 302-01-2	$H_2N\text{-}NH_2$	OH, O_3	<1	Ammonia, nitrogen	4,9,21,102,155, 170	
Hydrochloric acid (HCl) 7647-01-0	HCl	Deposition	1 to 5	Chloride salts (e.g., ammonium chloride)	133,134	
Hydrogen fluoride (HF) 7664-39-3	HF	Deposition	1 to 5	Fluoride salts	133,134	By analogy to HCl
Hydroquinone 123-31-9		OH	1 to 5	No information found	21	*
Isophorone 78-59-1		OH, NO_3, O_3	<1	No information found	21	*
Maleic anhydride 108-31-6	$O=C\text{-}CH=CHC=O$	OH	<1 to >5	Carbon dioxide, formic acid, carbon monoxide, aldehydes and esters which should photolyze	5,17,166	Significant disagreement over lifetime

TABLE 5.1 (CONTINUED)
Summary of Hazardous Air Pollutant Transformations and Lifetimes (Chemicals shown in italics are high priority urban HAPs)

Compound and CAS Number	Chemical Formula/Structure	Major Removal Processes	Reported Atmospheric Lifetimes (days)	Transformation Products	References	Comments/Notes
Methanol 67-56-1	CH_3OH	OH, deposition	1 to 5	Water, formaldehyde	4,16,17,19,21, 55,69	
Methoxychlor 72-43-5	CH_3O—(ring)—CH(H)(CCl_3)—(ring)—OCH_3	OH, deposition	<1	No information found	21	*
Methyl bromide (Bromomethane) 74-83-9	CH_3Br	OH	>5	Formaldehyde, formyl bromide, carbon monoxide, hydrogen bromide	14,21,37,40,50, 55,146	
Methyl chloride (Chloromethane) 74-87-3	CH_3Cl	OH	>5	Formaldehyde, formyl chloride, carbon monoxide, hydrogen chloride	14,21,17,37,50, 55,102	
Methyl chloroform (1,1,1-Trichloroethane) 71-55-6	CCl_3CH_3	OH	>5	Phosgene, formaldehyde, hydrogen chloride, carbon monoxide, acetyl chloride, chloroacetyl chloride, formyl chloride	5,6,21,55,129	
Methyl ethyl ketone (2-Butanone) 78-93-3	$C_2H_5COCH_3$	Photolysis, OH	<1 to >5	Carbon dioxide, carbon monoxide, water, formaldehyde, methanol, formic acid, acetic acid, acetaldehyde, PAN	8,17,18,21,22, 29,55,87,88, 166	Significant disagreement over lifetime

Compound / CAS	Structure	Process	Lifetime	Products	References
Methylhydrazine 60-34-4	CH_3NHNH_2	OH, O_3	<1	Amines	21,52,55,102, 122
Methyl iodide (Iodomethane) 74-88-4	CH_3I	Photolysis, OH	1 to 5	Formaldehyde, carbon monoxide, hydrogen iodide, formyl iodide	5,21,166
Methyl isobutyl ketone 108-10-1	$(CH_3)_2CHCH_2COCH_3$	OH	1 to 5	Acetone, PAN, formaldehyde	17,21,29,55
Methyl isocyanate 624-83-9	CH_3NCO	OH	<1	No information found	21
Methyl methacrylate 80-62-6	$H_2C=C(CH_3)CO_2CH_3$	OH, O_3	<1 to 1-5	Pyruvic acid, methyl pyruvate, epoxides	16,17,21
Methyl tert-butyl ether 1634-04-4	$(CH_3)_3COCH_3$	OH	1 to 5	Formaldehyde, acetone, methyl acetate, t-butyl formate	4,11,24,25,55, 156
4,4'-Methylenebis (2-chloroaniline) 101-14-4	* (structure: dichloro-diaminodiphenylmethane)	Photolysis, OH	<1	No information found	21
Methylene chloride 75-09-2	$Cl\text{-}CH_2\text{-}Cl$	*OH*	>5	*Formyl chloride, phosgene, carbon monoxide, carbon dioxide, hydrogen chloride*	*6,7,14,15,17,21, 69,102,129*
4,4'-Methylenediphenyl diisocyanate 101-68-8	* (structure: OCN–C₆H₄–CH₂–C₆H₄–NCO)	Deposition, OH	1 to 5	No information found	21,166
4,4'-Methylenedianiline 101-77-9	* (structure: H₂N–C₆H₄–CH₂–C₆H₄–NCO)	OH	<1	No information found	21,55,120

TABLE 5.1 (CONTINUED)
Summary of Hazardous Air Pollutant Transformations and Lifetimes (Chemicals shown in italics are high priority urban HAPs)

Compound and CAS Number	Chemical Formula/Structure	Major Removal Processes	Reported Atmospheric Lifetimes (days)	Transformation Products	References	Comments/Notes
Naphthalene 91-20-3	[structure: naphthalene]	OH, NO$_3$	<1	Methylnitronaphthalene, 3-nitrobiphenyl, 2-nitronaphthalene, 1-nitronaphthalene1- and 2-naphthols	8,9,17,19,21,45, 47,52,55,56,61, 68,78,90,105	
Nitrobenzene 98-95-3	[structure: nitrobenzene, NO$_2$]	OH	>5	Nitrophenols, dinitrobenzene, ring cleavage products	4,5,7,9,17,50,55, 102,126,157	
4-Nitrobiphenyl 92-93-3	[structure: 4-nitrobiphenyl, NO$_2$]	OH	1-5 to >5	No information found	21	*
4-Nitrophenol 100-02-7	$NO_2C_6H_4OH$	OH, photolysis	1-5 to >5	Dinitrophenols	19,21,117,146	Reaction of nitrophenols generally much slower than those of phenols (117)
2-Nitropropane 79-46-9	$CH_3CH(NO_2)CH_3$	Photolysis, OH	<1 to >5	Formaldehyde, acetaldehyde	5,21,171	Disagreement over lifetime
N-Nitroso-N-methylurea 684-93-5	$CH_3N(NO)C(O)NH_2$	OH, photolysis	<1	Aldehydes, nitramines	5,21	

Compound	Structure	Mechanism	Lifetime	Products	References	Products based on analogy with N-nitroso-diethylamine
N-nitrosodimethylamine 62-75-9	$(CH_3)_2NNO$	Photolysis, OH	<1	Nitric oxide, nitramines, aldehydes, methyl methyleneamine	5,21,52,55,107, 121,172	
N-Nitrosomorpholine 59-89-2		OH, photolysis	<1	Aldehydic ethers, nitric oxide	5,21	
Parathion 56-38-2		OH, O_3	<1	Paraoxon, 4-nitrophenol, diethylphosphoric acid, diethyl-thiophosphoric acid	21,33,95,97,135	
Pentachloronitrobenzene 82-68-8		Photolysis, deposition	>5	No information found	21	*
Pentachlorophenol 87-86-5		OH, photolysis	>5	No information found	21,23	
Phenol 108-95-2		OH,NO_3	<1	Dihydroxybenzenes, 2&4-nitrophenols, benzenetriols, glyoxal, glyoxalic acid	5,6,7,10,17,19, 21,50,51,54,55, 60,117	

TABLE 5.1 (CONTINUED)
Summary of Hazardous Air Pollutant Transformations and Lifetimes (Chemicals shown in italics are high priority urban HAPs)

Compound and CAS Number	Chemical Formula/Structure	Major Removal Processes	Reported Atmospheric Lifetimes (days)	Transformation Products	References	Comments/Notes
p-Phenylenediamine 106-50-3	NH_2 benzene ring NH_2	Photolysis, OH	<1	No information found	21	*
Phosgene 75-44-5	$COCl_2$	Liquid H_2O, deposition	1 to 5	Carbon dioxide, hydrogen chloride	5,6,17,21,40, 129,132	
Phosphine 7803-51-2	PH_3	OH	<1	Phosphate salts	19	
Phosphorus (elemental) 7723-14-0	P_4	O_2	<1	P_4O_6, P_4O_{10}, mixture of phosphorus acids (in presence of liquid water)	94,143	
Phthalic anhydride 85-44-9	(structure)	OH, deposition	>5	No information found	21	*
Polychlorinated biphenyls 1336-36-3	*Biphenyl with Cl at various levels of saturation*	*Photolysis, OH*	*>5*	*PCBs of lower chlorination*	*5,116,180, 181*	

Compound (CAS)	Structure	Reactive species	Lifetime (days)	Products	References	Notes
1,3-Propane sultone 1120-71-4	(structure)	OH	1 to 5	No information found	21	*
β-Propiolactone 57-57-8	(structure)	OH	>5	No information found	21	
Propionaldehyde 123-38-6	C_2H_5CHO	OH	<1	Acetaldehyde, formaldehyde, peroxypropionyl nitrate	4,21,69,166,168	See acetaldehyde for analogous products
Propoxur (Baygon) 114-26-1	(structure)	OH, deposition	<1 (gas phase) >5 (particle phase)	No information found	21,166	*
Propylene dichloride (1,2-Dichloropropane) 78-87-5	$CH_3CHClCH_2Cl$	*OH*	*>5*	*No information found*	*21,55,75*	*
Propylene oxide 75-56-9	CH_3CHCH_2 (structure)	OH	>5	Formaldehyde, glyoxylic acid, methylglyoxal, acetaldehyde, propionaldehyde, possibly acetone and PAN	5,6,7,17,55	
1,2-Propyleneimine (2-Methylaziridine) 75-55-8	CH_3CHCH_2 NH	OH	<1	No information found	21	*
Quinoline 91-22-5	(structure)	*OH*	*1 to 5*	*No information found*	*21*	*

TABLE 5.1 (CONTINUED)
Summary of Hazardous Air Pollutant Transformations and Lifetimes (Chemicals shown in italics are high priority urban HAPs)

Compound and CAS Number	Chemical Formula/Structure	Major Removal Processes	Reported Atmospheric Lifetimes (days)	Transformation Products	References	Comments/Notes
Quinone 106-51-4		OH	<1 to >5	No information found	21,167	* Significant disagreement over lifetime
Styrene 100-42-5	CH=CH$_2$	OH, O$_3$, NO$_3$	<1	Peroxybenzoylnitrate, benzaldehyde, formaldehyde, benzoic acid, formic acid	17,21,52,54,55, 71,81,103,144, 153,158,159	
Styrene oxide 96-09-3	CH—CH	OH	1 to 5	No information found	21	*
2,3,7,8-Tetrachloro-dibenzo-p-dioxin 1746-01-6		Photolysis, OH, deposition	1-5 to >5 (gas phase) >5 (particle phase)	Chlorinated dibenzo dioxins of lower chlorination	5,21,182-188	
1,1,2,2-Tetrachloroethane 79-34-5	Cl$_2$CHCHCl$_2$	OH	>5	No information found	21,168	*

Compound / CAS	Structure	Process	Half-life	Transformation products	References
Tetrachloroethylene (Perchloroethylene) 127-18-4	$Cl_2C=CCl_2$	OH	>5	Phosgene, formic acid, tri-chloroacetylchloride, carbon monoxide, hydrogen chloride	7,15,21,34,3,54, 55,69,80,129
Titanium tetrachloride 7550-45-0	$TiCl_4$	Deposition	>5	No information found	166
Toluene 108-88-3		OH	1 to 5	Benzaldehyde, nitrotoluene, o,m,p-cresols, nitrocresols, methyl nitrate, methyl glyoxal, PAN, formaldehyde, benzyl-nitrate, acetaldehyde, glyoxal, dicarbonyls, benzoic acid, 2-butenoic acid, 2-butenal, acetic acid, butenedial, 4-oxo-2-pentenal, 4-oxo-2-butenoic acid, 4-oxo-2-pentenoic acid, glycolaldehyde, hydroxyacetone, benzo-quinone, methyl benzo-quinone, 2-5-furandione 3-methyl-2,5-furandione dihydro-2,5-furandione	4,5,6,7,9,14,17 ,19,21,35,52,54 ,55,65,102,110, 138,140,141, 164,179,189, 190
Toluene-2,4-diamine 95-80-7		OH	<1	No information found	21,55,102,120
2,4-Toluene diisocyanate 584-84-9		OH, photolysis, liquid water	<1	Toluenediamine and diurethanes	21,55,120,132, 142

TABLE 5.1 (CONTINUED)
Summary of Hazardous Air Pollutant Transformations and Lifetimes (Chemicals shown in italics are high priority urban HAPs)

Compound and CAS Number	Chemical Formula/Structure	Major Removal Processes	Reported Atmospheric Lifetimes (days)	Transformation Products	References	Comments/Notes
o-Toluidine 95-53-4		OH	<1	No information found	21	*
Toxaphene 8001-35-2		Deposition	>5	No information found	96	
1,2,4-Trichlorobenzene 120-82-1		OH	>5	No information found	19,21,23,51,55, 56,72,82,89, 102	
1,1,2-Trichloroethane 79-00-5	CH$_2$ClCHCl$_2$	OH	>5	No information found	21,55,84,85,102	

Compound	Structure	Oxidant	Lifetime	Products	References
Trichloroethylene 79-01-6	$Cl_2C=CHCl$	OH	>5	*Phosgene, dichloroacetyl-chloride, formyl chloride, trichloroethylene oxide, chloroform, formic acid, carbon monoxide, hydrogen chloride*	*4–7,13,15,17,21, 22,34,35,69, 129*
2,4,5-Trichlorophenol 95-95-4		OH	>5	No information found	10,12,21,23
2,4,6-Trichlorophenol 88-06-2		OH	>5	No information found	10,21,23
Triethylamine 121-44-8	$(C_2H_5)_3N$	OH, HNO_3	<1	No information found	168
Trifluralin 1582-09-8		Photolysis, OH, O, O_3	<1	Dealkylated trifluralin	97,98
2,2,4-Trimethylpentane 540-84-1	$CH_3C(CH_3)_2CH_2CH\text{-}(CH_3)CH_3$	OH	1 to 5	Aldehydes, ketones	55,102,115

*

TABLE 5.1 (CONTINUED)
Summary of Hazardous Air Pollutant Transformations and Lifetimes (Chemicals shown in italics are high priority urban HAPs)

Compound and CAS Number	Chemical Formula/Structure	Major Removal Processes	Reported Atmospheric Lifetimes (days)	Transformation Products	References	Comments/Notes
Vinyl acetate 108-05-4	$CH_3CO_2CH=CH_2$	OH	<1	Formaldehyde	17,166,173	
Vinyl bromide 593-60-2	$H_2C=CHBr$	OH, O_3	<1 to 1-5	Formaldehyde, formyl bromide	4,21,55,57,102, 166	See vinyl chloride for analogous products
Vinyl chloride 75-01-4	*$H_2C=CHCl$*	*OH, O_3*	*<1 to 1-5*	*Formic acid, formyl chloride, hydrogen chloride, carbon monoxide, formaldehyde, carbon dioxide, monochloro-acetaldehyde, monochloro-peroxyacetyl nitrate*	4,21,34,38,50, 52,54,55,57–59	
Vinylidene chloride (1,1-Dichloroethylene) 75-35-4	$H_2C=CCl_2$	OH, O_3	<1	Phosgene, formaldehyde, formic acid, chloroacetyl chloride, carbon monoxide	7,13,21,38,52, 54,55,74,77,83, 93	
Xylenes (mixed) 1330-20-7	⬡ + $(CH_3)_2$	OH	<1	Substituted benzaldehydes, hydroxy xylenes, nitro compounds, methyl glyoxal, biacetyl, ring cleavage products	5,21,35	

Compound	Structure	Reaction	Lifetime	Transformation products	References
o-Xylene 95-47-6	![CH3 CH3 benzene structure]	OH	<1	Tolualdehyde, dimethyl phenols, dimethyl nitro-benzene, maleic anhydride, biacetyl, glyoxal, methyl glyoxal, acetaldehyde, methyl benzoquinone, 2,3-butenedione, glycolaldehyde, hydroxyacetone formaldehyde, CO, PAN, butenedial, 4-oxo-2-pentenal, 3-methyl-4-oxo-2-pentenal, carboxylic acids	8,17,19,44,52, 54,55,60,65, 103,110,160, 161,164,189
m-Xylene 108-38-3	![CH3 CH3 benzene structure]	OH	<1	Tolualdehyde, formaldehyde, methyl glyoxal, acetalde-hyde, aliphatic dicarbonyls, PAN, CO, glyoxal, 4-oxo-2-pentenal, 2-methyl-2-butenedial, 2-methyl-4-oxo-2-pentenal, carboxylic acids, dimethyl phenols, methylbenzoquinone, glycolaldehyde, hydroxy-acetone, m-toluic acid, 3-methyl-2,5-furandione	8,9,14,17,21,55, 103,110,160, 164,179,189
p-Xylene 106-42-3	![CH3 CH3 benzene structure]	OH	<1 to 1-5	Tolualdehyde, methyl glyoxal, glyoxal, dimethyl phenol, formic acid, PAN, cis-3-hexene-2,5-dione, formaldehyde, dimethyl-nitrobenzene, methylbenzyl nitrate, 2,5-dimethyl-2,4-hexadienedial, dimethyl-benzoquinone, benzo-quinone, glycoaldehyde, hydroxyacetpme, m-methyl butenedial	6,8,17,19,21,41, 52,54,55, 61–67,110,160, 161,164,189
Antimony Compounds	Inorganic compounds in particle phase	Deposition	>5	No significant chemical transformations	5
Arsenic Compounds	*Primarily inorganic arsenic compounds in particle phase*	*Deposition*	*>5*	*No significant chemical transformations*	*5*

TABLE 5.1 (CONTINUED)
Summary of Hazardous Air Pollutant Transformations and Lifetimes (Chemicals shown in italics are high priority urban HAPs)

Compound and CAS Number	Chemical Formula/Structure	Major Removal Processes	Reported Atmospheric Lifetimes (days)	Transformation Products	References	Comments/Notes
Beryllium Compounds	*Inorganic compounds in particle phase*	*Deposition*	*>5*	*No significant chemical transformations*	*5*	
Cadmium Compounds	*Inorganic compounds in particle phase*	*Deposition*	*>5*	*No significant chemical transformations*	*5*	
Chromium Compounds	*Inorganic compounds in particle phase*	*Deposition*	*>5*	*No significant chemical transformations*	*5*	
Cobalt Compounds	Inorganic compounds in particle phase	Deposition	>5	No significant chemical transformations	5	
Coke Oven Emissions	*Mixture of organic and inorganic vapors and particles*	*No information found*	*No information found*	*No information found*		*Species undefined (see polycyclic organic matter)*
Cyanide Compounds	e.g., HCN (74-90-8), propionitrile, C_2H_5CN (107-12-0), cyanogen, C_2N_2 (74-87-5) (See also acetonitrile, acrylonitrile)	OH, O_3	>5	No information found	9,21,52,54,55, 69,127	
Glycol Ethers	e.g., $HOCH_2CH_2OCH_3$	OH	<1	Hydroxy esters, formaldehyde, peroxyacyl nitrates, hydroxyacids, hydroxycarbonyls	17,55,146,154	

Lead Compounds	Primarily inorganic compounds in particle phase.	Deposition	>5	No significant chemical transformations	5
	Trace amounts of organo-lead compounds, e.g., tetraethyl lead $(C_2H_5)_4Pb$ (78-00-2)	OH, O_3 photolysis	<1 to 1-5	Inorganic Pb compounds	21,52,55,123,124
Manganese Compounds	Inorganic compounds in the particle phase	Deposition	>5	No significant chemical transformations	5
Mercury Compounds	Primarily elemental mercury vapor, Hg°	Aqueous OH, aqueous O_3, O_3 deposition, Cl species	>5	Particulate Hg, reactive gaseous Hg (see below)	104,174-178
	Particulate inorganic Hg	Deposition	>5	No significant chemical transformations	5
	Reactive gaseous Hg, primarily Hg (II)	Aqueous HO_2, photoreduction in solution, deposition	<1	Hg°, particulate Hg compounds (see above)	55,125,176,177
Fine Mineral Fibers	Fibrous glass, mineral wool, or ceramic fibers, similar in shape but chemically and physically distinct from asbestos fibers	Deposition	>5	No chemical transformations	112
Nickel Compounds	Inorganic compounds in the particle phase	Deposition	>5	No significant chemical transformations	5
Polycyclic Organic Matter (see text)	e.g., Phenanthrene (85-01-8), anthracene (120-12-7), (See also naphthalene)	OH, NO_2	<1 to 1-5	Nitroarenes, nitronaphthalenes, nitrofluoranthenes, nitropyrenes, nitroacenaphthalenes	45,46,47,105,118,139

TABLE 5.1 (CONTINUED)
Summary of Hazardous Air Pollutant Transformations and Lifetimes (Chemicals shown in italics are high priority urban HAPs)

Compound and CAS Number	Chemical Formula/Structure	Major Removal Processes	Reported Atmospheric Lifetimes (days)	Transformation Products	References	Comments/Notes
Radionuclides (including Radon)	Radon gas	Radioactive decay	1 to 5	Radioactive daughters in the particle phase (see below): $^{218}Po, ^{214}Pb, ^{214}Bi, ^{214}Po)$	113,114	
	Radioactive compounds in the particle phase	Deposition	>5	No significant chemical transformations		
Selenium Compounds	Inorganic compounds in the particle phase	Deposition	>5	No significant chemical transformation	5	

* Atmospheric lifetime calculated using rate constants estimated from structure-activity relationships.

References

1. Atkinson, R., Comments on predicting gas phase organic molecule reaction rates using linear free energy correlations, I. O(^{3}P) and OH addition and abstraction reactions, *Int. J. Chem. Kinet.* 12: 761, 1980.

2. Atkinson, R. et al., Rate constants for the gas phase reactions of O_3 with a series of carbonyls at 296 K. Funded under toxics CCULIA Task 0518, Statewide Air Pollution Research Center, University of California, Riverside, CA, 1981.

3. Pitts, J.N. Jr. et al., Experimental protocol for determining absorption cross sections of organic compounds, EPA-600/3-81-051, U.S. EPA, Research Triangle Park, NC, 1981.

4. Atkinson, R. et al., Atmospheric fates of organic chemicals: Prediction of ozone and hydroxyl radical reaction rates and mechanisms, EPA-600/3-85-063, U.S. EPA, Research Triangle Park, NC, 1985.

5. Cupitt, L.T. Fate of toxic and hazardous materials in the air environment, EPA-600/53-80-084, U.S. EPA, Research Triangle Park, NC, 1980.

6. Anderson, G.E. Human exposure to atmospheric concentrations of selected chemicals, EPA report #SAI/EFB1-156R2, U.S. EPA, Research Triangle Park, NC, 1983, NTIS Publication No. PB83-265249.

7. Spicer, C. W., Riggin, R. M., Holdren, M. W., DeRoos, F. L. and Lee, R. N., Atmospheric reaction products from hazardous air pollutant degradation. EPA Contract No. 68-02-3169 (WA-33/40), U.S. EPA, Research Triangle Park, NC, 1984.

8. Klöpffer, W., Kaufmann, G. and Frank, R., Phototransformation of air pollutants: Rapid test for the determination of K_{OH}. *Z. Naturforsch*, 40a: 686, 1985.

9. Pitts, J.N. Jr. et al., Formation and fate of toxic chemicals in CA's atmosphere, Final Report, Contract No. A3-126-32 CA Air Resources Board, Sacramento, CA, 1985.

10. Leuenberger, C., Ligocki, M. P. and Pankow, J. F., Trace organic compounds in rain. 4. Identities, concentrating and scavenging mechanisms for phenols in urban air and rain, *Environ. Sci. Technol.*, Vol. 19, No. 11: 1053, 1985.

11. Munshi, H.B., et al., Rate constants of the reactions of ozone with nitriles, acrylates and terpenes in gas phase, *Atmospheric Environment*, Vol. 23, No. 9: 1971, 1989.

12. Bunce, N.J. and Nakai, J.S., Atmospheric chemistry of chlorinated phenols, *J. Air Poll. Control Assoc.*, 39: 820, 1989.

13. Edney, E., Mitchell, S. and Bufalini, J.J., Atmospheric chemistry of several toxic compounds, COO/01-CCULIA U.S. EPA, Research Triangle Park, NC, 1982.

14. Altshuller, A.P., Lifetimes of organic molecules in the troposphere and lower stratosphere, In: J.N. Pitts, Jr. and R.L. Metcalf (Eds.), *Advances in Environmental Science and Technology*, Vol. 10, 1980, p. 181. John Wiley & Sons.

15. Cupitt, L.T., Atmospheric persistence of eight air toxics, EPA-600/3-87-004, U.S. EPA, Research Triangle Park, NC, 1987.

16. Atkinson, R., Carter, W.P.L. and Winer, A.M., Evaluation of hydrocarbon reactivities for use in control strategies, Final Report, Contract No. AO-105-32, CA Air Resources Board, Sacramento, CA, 1983.

17. Singh, H.B., Jaber, H.M. and Davenport, J.E., Reactivity/volatility classification of selected organic chemicals: existing data, EPA-600/3-84-082, U.S. EPA, Research Triangle Park, NC, 1984.

18. U.S. EPA, Health effects assessment for methyl ethyl ketone, ECAO-CIN-H003. Environmental Criteria and Assessment Office, Cincinnati, OH, 1984.

19. OH Reaction rate constants and trospheric lifetime of selected environmental chemicals, Report 1980-1983, In: K.H. Becher et al. (Eds.), *Methods of the Ecotoxicological Evaluation of Chemicals, Photochemical Degradation in the Gas Phase*, Vol. 6, 1984.

20. Finlayson-Pitts, B.J. and Pitts, J.N. Jr. et al., The chemical basis of air quality: Kinetics and mechanisms of photochemical air pollution and application to control strategies, In: J.N. Pitts, Jr. and R.L. Metcalf (Eds.), *Advances in Environmental Science and Technology*, Vol. 7, 75, 1977.

21. Howard, P.H et al., *Handbook of Environmental Degradation Rates*, Lewis, 1991.

22. Raber, W., Reinholdt, K. and Moortgat, G.K., Photooxidation study of methylethylketone and methylvinylketone. In: *Proc. 5th Euro. Symp. on Physics-Chemical Behavior of Atmospheric Pollutants*, Varese, Italy, 364, 1989.

23. Bunce, N.J., Nakai, J.S. and Yawching, M., Estimates of the tropospheric lifetimes of short- and long-lived atmospheric pollutants, *J. Photochem. Photobiol.*, A: Chem. 57: 429, 1991.

24. Wallington, T.J. et al., Gas-phase reactions of hydroxyl radicals with the fuel additives methyl *tert*-butyl ether and *tert*-butyl alcohol over the temperature range 240–440 K, *Environ. Sci. Technol.*, 22, 842, 1988.

25. Wallington, T.J. et al., Kinetics of the reaction of OH radicals with a series of ethers under simulated atmospheric conditions at 295 K, *Int. J. Chem. Kinet.*, 21, 993, 1989.

26. Kurylo, M.J., Elementary reactions of atmospheric sulfides, *J. Photochem.*, 9: No. 2/3: 124, 1978.

27. Atkinson, R., Perry, R.A. and Pitts, J.N. Jr., Rate constants for the reaction of OH radicals with COS, CS_2 and CH_3SCH_3 over the temperature range 299-430 K, *Chem. Phys. Letters.*, 54, No. 1, 14, 1978.

28. Dagout, P. et al., Flash photolysis resonance fluorescence investigation of the gas-phase reactions of hydroxyl radicals with cyclic ethers, *J. Phys. Chem.*, 94, 1881, 1990.

29. Cox, R.A., Patrick, K.F. and Chant, S.A., Mechanism of atmospheric photooxidation of organic compounds. Reactions of alkoxy radicals in oxidation of n-butane and simple ketones, *Environ. Sci. Tech.*, 15, 587 1981.

30. Atkinson, R., Estimation of OH radical reaction rate constants and atmospheric lifetimes for polychlorobiphenyls, dibenzo-p-dioxins and dibenzofurans, *Environ. Sci. Technol.*, 21, 305, 1987.

31. Atkinson, R., Aschmann, S.M. and Winer, A.M., Kinetics of the reactions of NO_3 radicals with a series of aromatic compounds, *Environ. Sci. Technol.*, 21, 1123, 1987.

32. Tuazon, E.C. et al., Atmospheric loss process of 1,2-dibromo-3-chloropane and trimethylphosphate, *Eviron. Sci. Technol.*, 20 (10), 1043, 1986.

33. Plimmer, J.R. and Johnson, W. E., Pesticide degradation products in the atmosphere, In: L. Somasundaram and J.R. Coats (Eds.), *Pesticide Transformation Products, Fate and Significance in the Environment*, ACS Symposium Series 459, American Chemical Society, 274, 1991.

TABLE 5.1 (CONTINUED)
Summary of Hazardous Air Pollutant Transformations and Lifetime

34. Pitts, J.N. Jr. et al., Formation and fate of toxic chemicals in California's atmosphere, Final Report, Contract No. A2-115-32, CA Air Resources Board, Sacramento, CA, 1984.

35. Pitts, J.N. Jr. et al., Chemical consequences of air quality standards and of control implementation programs, Final Report, Contract No. A1-030-32, CA Air Resources Board, Sacramento, CA, 1983.

36. Atkinson, R. et al., Rate constants for the reaction of OH radicals with a series of alkanes and alkenes at 299 ± 2 K, Statewide Air Pollution Research Center, Riverside, CA, Funded under Toxics MOO/02/CCULIA, Task 0518, 1981.

37. Weller, R., Lorenzer-Schmidt, H. and Schrems, O., FTIR studies on the photooxidation mechanisms of methyl chloride, methyl bromide, bromoform and bromotrifluoromethane, *Ber. Bunsen-Ges. Phys. Chem.*, 93(3), 409, 1992.

38. Gay, B.W. Jr. et al., Atmospheric oxidation of chlorinated ethylenes, *Environ. Sci. Technol.*, 10(1), 58, 1976.

39. Atkinson, R., Aschmann, S.M. and Goodman, M.A., Kinetics of the gas-phase reactions of nitrate radicals with a series of alkynes, haloalkenes and alpha and beta-unsaturated aldehydes, *Int. J. Chem. Kinet.*, 19(4), 299, 1987.

40. Grosjean, D., Atmospheric chemistry of toxic contaminants, 4. Saturated halogenated aliphatics: methyl bromide, epichlorohydrin, phosgene, *J. Air. Waste Manage. Assoc.*, 41, 56, 1991.

41. Becker, K.H. and Klein, T., OH - initiated oxidation of p-xylene under atmospheric conditions, In: *Proc. 4th Euro. Symp. Physico-Chemical Behavior of Atmospheric Pollutants*, Stresa, Italy, 320, 1986.

42. Edney, E.O. and Corse, E.W., Hydroxyl radical rate constant intercomparison study, Report from Northrop Services, Inc., to U.S. EPA under Contract No. 68-02-4033, EPA/6003/3-86/056, 1986.

43. Perry, R.A., Atkinson,R. and Pitts, J.N. Jr. Kinetics and mechanism of the gas-phase reaction of OH radicals with methoxybenzene and o-cresol over the temperature range 299-435K, *J. Phys. Chem.*, 81(17), 1607, 1977.

44. Casado, J. Herrman, J.M. and Pichat, P., Phototransformation of o-xylene over atmospheric solid aerosols in the presence of O$_2$ and H$_2$O, In: *Proc. 5th Euro. Symp. Physico-Chemical Behavior of Atmospheric Pollutants*, Varese, Italy, 283, 1989.

45. Arey, J. et al., Nitroarene products from the gas-phase reactions of volatile polycyclic aromatic hydrocarbons with the hydroxyl radical and dinitrogen pentoxide, *Int. J. Chem. Kinet.*, 21(9), 775, 1989.

46. Atkinson, R. et al., Kinetics and nitroproducts of the gas-phase hydroxyl and nitrogen oxide (NO$_3$) radical-initiated reactions of naphthalene-d8, fluoranthene-d10 and pyrene, *Int. J. Chem. Kinet.*, 22(9), 999, 1990.

47. Atkinson, R et al., Kinetics and products of the gas-phase reactions of OH radicals and N$_2$O$_5$ with naphthalene and biphenyl, *Environ. Sci. Technol.*, 21(10), 1014, 1987.

48. Atkinson, R., Estimation of OH radical reaction rate constants and atmospheric lifetimes for polychlorobiphenyls, dibenzo-p-dioxins and dibenzofurans, *Environ. Sci. Technol.*, 21(3), 305, 1987.

49. Atkinson, R. and Aschmann, S.M., Rate constants for the gas-phase reaction of hydroxyl radicals with biphenyl and monochlorobiphenyls at 295 ± 1 K, *Environ. Sci. Technol.*, 19(5), 462, 1985.

50. Atkinson, R., Kinetics and mechanisms of the gas-phase reactions of the hydroxyl radical with organic compounds under atmospheric conditions, *Chem. Rev.*, 86: 69, 1986.

51. Becker, K.H. et al., Methods of the ecotoxicological evaluation of chemicals. Photochemical degradation in the gas phase, Vol. 6: OH reaction rate constants and tropospheric lifetimes of selected environmental chemicals, Report 1980-1983, Kernforschungsonlage Jülich GmBH, Projektträgerschaft Umweltshemikalien, Jül-Spez, 279, 1984. ISSN 0343-7639.

52. Atkinson, R. and Carter, W.P.L., Kinetics and mechanics of the gas-phase reactions of ozone with organic compounds under atmospheric conditions, *Chem. Rev.*, 84:.437, 1984.

53. Warneck, P., *Chemistry of the Natural Atmosphere*, International Geophysics Series. Vol. 41, Academic Press, San Diego, 1988.

54. Atkinson, R., Kinetics and mechanisms of the gas-phase reactions of the nitrate (NO_3) radical with organic compounds, *J. Phys. Chem.*, Ref. Data 20(3), 459, 1991.

55. Atkinson, R., Kinetics and mechanism of the gas-phase reactions of the hydroxyl radical with organic compounds, *J. Phys. Chem.*, Ref. Data, Monograph 1, Am. Inst. Physics, New York, 1989.

56. Klöpffer, W. et al., Testing of the abiotic degradation of chemicals in the atmosphere: the smog chamber approach, *Ecotox. Environ. Safety*, 15: 298, 1988.

57. Perry, R.A., Atkinson, R. and Pitts, J.N. Jr., Rate constants for the reaction of OH radicals with $CH_2 = CF$, $CH_2 = CHCl$ and $CH_2 = CHBr$ over the temperature range 299-426 K, *J. Chem. Phys.*, 67, 458, 1977.

58. Gay, B.W. Jr. et al., Atmospheric oxidation of chlorinated ethylenes, *Environ. Sci. Technol.*, 10, 58, 1976.

59. Zhang, J., Hatakeyama, S. and Akimoto, H., Rate constants of the reaction of ozone with trans-1,2-dichloroethene and vinylchloride, *Int. J. Chem. Kinet.*, 15, 655, 1983.

60. Atkinson, R. et al., Kinetics of the gas phase reactions of NO_3 radicals with a series of aromatics at 296 ± 2 K, *Int. J. Chem. Kinet.*, 16, 887, 1984.

61. Atkinson, R. and Aschmann, S.M., Rate constants for the gas-phase reactions of the OH radical with a series of aromatic hydrocarbons at 296 ± 2 K, *Int. J. Chem. Kinet.*, 21, 355, 1989.

62. Doyle, G.J. et al., Gas-phase kinetic study of relative rates of reaction of selected aromatic compounds with hydroxyl radicals in an environmental chamber, *Environ. Sci. Technol.*, 9, 237, 1975.

63. Hansen, D.A., Atkinson, R. and Pitts, J.N. Jr. Rate constants for the reaction of OH radicals with a series of aromatic hydrocarbons, *J. Phys. Chem.*, 79, 1763, 1975.

64. Perry, R.A., Atkinson, R. and Pitts, J.N. Jr., Kinetics and mechanism of the gas-phase reaction of OH radicals with aromatic hydrocarbons over the temperature range 296-473 K, *J. Phys. Chem.*, 81, 296, 1977.

65. Pate, C.T., Atkinson, R. and Pitts, J.N. Jr., The gas-phase reaction of ozone with a series of aromatic hydrocarbons, *J. Environ. Sci. Health*, Part A, 11, 1, 1976.

66. Ravishankara, A.R. et al., A kinetic study of the reactions of OH with several aromatic and olefinic compounds, *Int. J. Chem. Kinet.*, 10, 783, 1978.

67. Nicovich, J.M., Thompon, R.L. and Ravishankara, A.R., Kinetics of the reactions of the hydroxyl radical with xylenes, *J. Phys. Chem.*, 85, 2913, 1981.

68. Atkinson, R. and Aschmann, S.M., Kinetics of the reaction of naphthalene, 2-methylnaphthalene and 2,3-dimethyl-naphthalene with OH radicals and with O_3 at 295 ± 1 K, *Int. J. Chem. Kinet.*, 18, 569, 1986.

69. Atkinson, R. et al., Evaluated kinetic and photochemical data for atmospheric chemistry, Supplement III IVPAC Subcommittee on gas kinetic data evaluation for atmospheric chemistry, *J. Phys. Chem.* Ref. Data, 18, 881, 1989.

70. Wahner, A. and Zetzach, C., Rate constants for the addition of OH to aromatics (benzene, p-chloroaniline and o-,m- and p-dichlorobenzene) and the unimolecular decay of the adduct, Kinetics into a quasi equilibrium, *Int. J. Phys. Chem.*, 87, 4945, 1983.

71. Atkinson, R. and Aschmann, S.M., Kinetics of the reaction of acenaphthene and acenaphthylene and structurally related aromatic compounds with OH and NO_3 radicals, N_2O_5 and O_3 at 296 ± 2 K, *Int. J. Chem. Kinet.*, 20, 513, 1988.

72. Klöpffer, W. et al., Quantitative erfassung der photochemischen transformationsprozesse in der troposphere, *Chemiker-Zeitung*, 110, 57, 1986.

73. Atkinson, R. et al., Kinetics and mechanisms of the reaction of the hydroxyl radical with organic compounds in the gas phase, *Adv. Photochem.*, 1979, Vol. 11, John Wiley & Sons, New York, pg. 375.

74. Tuazon, E.C. et al., Atmospheric reactions of chlorethenes with the OH radical, *Int. J. Chem. Kinetics*, 20, 241, 1988.

75. Tuazon, E.C. et al., A study of the atmospheric reactions of 1,3 dichloropropene and other selected organochlorine compounds, *Arch. Environ. Toxicol.*, 13, 691, 1984.

TABLE 5.1 (CONTINUED)
Summary of Hazardous Air Pollutant Transformations and Lifetime

76. Howard, C.J. and Evenson, K.M., Rate constants for the reactions of OH with ethane and some halogen substituted ethanes at 296 K, *J. Chem. Phys.*, 64, 4303, 1976.

77. Hull, L.A., Hisatsune, I.C. and Heicklen, J., The reaction of O_3 with CCl_2CH_2, *Can. J. Chem.*, 51, 1504, 1973.

78. Atkinson, R., Aschmann, S.M. and Pitts, J.N. Jr., Kinetics of the reactions of naphthalene and biphenyl with OH radicals and with O_3 at 294 ± 1 K, *Environ. Sci. Technol.*, 18, 110, 1984.

79. Atkinson, R. et al., Rate constants for the gas-phase reactions of ozone with a series of carbonyls at 296 K, *Int. J. Chem. Kinet.* 13, 1133, 1981.

80. Atkinson, R. et al., Kinetics and atmospheric implications of the gas-phase reactions of NO_3 radicals with a series of monoterpenes and related organics at 294 ± 2 K, *Environ. Sci. Technol.*, 19, 159, 1985.

81. Bufalini, J.J. and Altshuller, A.P., Kinetics of vapor-phase hydrocarbon–ozone reactions, *Can. J. Chem.*, 43, 2243, 1965.

82. Rinke, M. and Zetsch, C., Rate constants for the reaction of OH radicals with aromatics: benzene, phenol, aniline and 1,2,4-trichlorobenzene. *Ber. Bunsen-Ges. Phys. Chem.*, 88, 55, 1984.

83. Edney, E.O., Kleindienst, T.E. and Corse, E.W., Room temperature rate constants for the reaction of OH with selected chlorinated and oxygenated hydrocarbons, *Int. J. Chem. Kinet.*, 18, 1355, 1986.

84. Jeong, K. et al., Kinetics and reactions of OH with C_2H_2, CH_3CCl_3, CH_3CCl, $CH_2ClCHCl_2$, $CH_2ClCClF_2$ and CH_2FCF_3, *J. Phys. Chem.*, 88, 1222, 1984.

85. Jeong, K. and Kaufmann, F., Rates of reactions of 1,1,1-trichloroethane (methylchloroform) and 1,1,2-trichloroethane with OH., *Geophys. Res. Lett.*, 6, 757, 1979.

86. Wallington, T.S., Dagout, P. and Kurylo, M.J., Correlation between gas-phase and solution-phase reactivities of hydroxyl radicals toward saturated organic compounds, *J. Phys. Chem.*, 92, 5024, 1988.

87. Carlier, P., Hanachi, H. and Mouvier, G., The chemistry of carbonyl groups in the atmosphere: a review, *Atmos. Environ.*, 20, 2079, 1986.

88. Demerjian, K.L., Kerr, J.A. and Calvert, J.G., The mechanisms of photochemical smog formation, In: Pitts, J.N. Jr. and Metcalf, R.L., (Eds.), *Advances in Environmental Science and Technology*, John Wiley & Sons, Vol. 4, 1974, p.1.

89. Bunce, N.J. et al., An assessment of the importance of direct solar degradation on some simple chlorinated benzenes and diphenyls in the vapor phase, *Environ. Sci. Technol.*, 23, 213, 1989.

90. Lorenz, K. and Zeller, R., Kinetics of reactions of OH-radicals with benzene, benzene-d6 and naphathalene. *Ber. Bunsen-Ges. Phys. Chem.*, 87, 629, 1983.

91. Seinfeld, J.H., *Atmospheric Chemistry and Physics of Air Pollution*, Wiley, New York, 1986.

92. Atkinson, R. et al., Kinetics of the gas-phase reactions of NO_3 radicals with a series of alkanes, *J. Phys. Chem.*, 88, 2361, 1988.

93. Atkinson, R., Aschmann, S.M. and Goodman, M.A., Kinetics of the gas-phase reactions of NO_3 radicals with a series of alkynes, haloalkenes and α,β-unsaturated aldehydes, *Int. J. Chem. Kinet.*, 19, 299, 1987.

94. Cotton, F.A. and Wilkinson, G., *Advanced Inorganic Chemistry*, 2nd ed., Interscience Publishers, 1966, pp 489-518.

95. Winer, A.M. and Atkinson, R., Atmospheric reaction pathways and lifetimes for organophosphorus compounds, In: D.A. Kurtz (Ed.), *Long Range Transport of Pesticides*, Lewis: Chelsea, MI, 115, 1990.

96. Voldner, E.C. and Schroeder, W.H., Modeling of atmospheric transport and deposition of toxaphene into the Great Lakes ecosystem, *Atmos. Environ.*, 23(9), 1949, 1989.

97. Woodrow, J.E. et al., Rates of transformation of trifluralin and parathion vapors in air, *J. Agric. Food Chem.*, 26(6), 1312, 1978.

98. Mongar, K. and Miller G.C., Vapor phase photolysis of trifluralin in an outdoor chamber, *Chemosphere*, 17(11), 2183, 1988.

99. Japar, S.M. et al., Atmospheric reactivity of gaseous dimethyl sulfate, *Environ. Sci. Technol.*, 24, 313, 1990.

100. Japar, S.M. et al., Atmospheric chemistry of gaseous diethyl sulfate, *Environ. Sci. Technol.*, 24, 894, 1990.

101. Betterton, E.A., Henry's law constants of soluble and moderately soluble organic gases: effects on aqueous phase chemistry, In: J.O. Nriagu (Ed.), Gaseous Pollutants: Characterization and Cycling, John Wiley & Sons, New York, 1, 1992.

102. Leifer, A., Determination of rates of reaction in the gas-phase in the troposphere. Theory and practice, 3. Rates of indirect photoreaction: Technical support document for test guideline 796.3900, EPA-700\R-92-002, U.S. EPA, Research Triangle Park, NC, 221 pp, 1992.

103. Atkinson, R et al., Rate constants for the gas-phase reactions of O₃ with selected organics at 296 K, *Int. J. Chem. Kinet.*, 14, 13, 1982.

104. Petersen, G., et al., Numerical modelling of the atmospheric transport, chemical transformations and deposition of mercury, presented at 18th Int Tech. Mtg. on Air Pollution Modelling and its Applications, sponsored by NATO-CCMS, Vancouver, BC, May 13-17, 1990. (Printed as report number GKSS-90E24 by the GKSS-Forschungszentrum Geesthach GmbH, Hamburg, Germany.)

105. Pitts, J.N. Jr., Nitration of gaseous polycyclic aromatic hydrocarbons in simulated and ambient urban atmospheres: a source of mutagenic nitroarenes, *Atmos. Environ.*, 21, 2531, 1987.

106. Pitts, J.N. Jr. et al., Chemical consequences of air quality standards and control implementation programs, Final report to the CA Air Resources Board on Contract Number A8-145-31, prepared by the Statewide Air Pollution Research Center, University of CA-Riverside, NTIS No. PB84-111178, March 1981.

107. Carter, W.P.L. et al., Experimental protocol for determining photolysis reaction rate constants, Final report on Cooperative Agreement Number CR806661-01, EPA-600\3-83-100, prepared for U.S. EPA by the Statewide Air Pollution Research Center, University of California Riverside, October 1983.

108. Grosjean, D., Atmospheric chemistry of toxic contaminants. 3. Unsaturated aliphatics: acrolein, acrylonitrile, maleic anhydride, *J. Air Waste Manage. Assoc.*, 40, 1664, 1990.

109. Tuazon, E.C. et al., Gas-phase reaction of 1,1-dimethylhydrazine with nitrogen dioxide, *J. Phys. Chem.*, 87, 1600, 1983.

110. Spicer, C.W. et al., Products from the atmospheric reactions of a series of aromatic hydrocarbons: toluene, o-xylene, m-xylene, p-xylene and ethylbenzene, Battelle final report to US EPA, Cooperative Agreement No. 810276, Columbus, OH, June, 1985.

111. Hov, O., The effect of chlorine on the formation of photochemical oxidants in southern Telemark, Norway, *Atmos. Environ.*, 19, 471, 1985.

112. Asbestiform Fibers: *Nonoccupational Health Risks*, National Academy Press, National Academy of Sciences, Washington, D.C., 1984.

113. Exposures from the uranium series with emphasis on radon and its daughters, NCRP Report No. 77, National Council on Radiation Protection and Measurements, Bethesda, MD, 1984.

114. Evaluation of occupational and environmental exposures to radon and radon daughters in the United States, NCRP Report No. 78, National Council on Radiation and Measurements, Bethesda, MD, 1984.

115. Grosjean, D. and Seinfeld, J.H., Parameterization of the formation potential of secondary organic aerosols, *Atmos. Environ.*, 23, No. 8: 1733, 1989.

116. Manchester-Neesvig, J.B. and Andren, A.W., Seasonal variation in the atmospheric concentration of polychlorinated biphenyl congeners, *Environ. Sci. Technol.*, 23, 1138, 1989.

117. Atkinson, R., Aschmann, S.M. and Arey, J., Reactions of OH and NO₃ radicals with phenol, cresols and 2-nitrophenol at 296 ± 2K, *Environ. Sci. Technol.*, 26: 1397, 1992.

118. Greenberg, A. et al., Fate of airborne PAH and nitroPAH, paper presented at 1987 meeting of the Air Pollution Control Association, New York, June 21-26, 1987.

119. Baulch, D.L. et al., Evaluated kinetic and photochemical data for atmospheric chemistry: Supplement II. CODATA Task Group on Gas Phase Chemical Kinetics, *J. Phys. Chem. Ref. Data*, 13, 1259, 1984.

120. Becker, K.H., Bastian, V. and Klein, T., The reaction of OH radicals with toluenediisocyanate, toluenediamine and methylenedianiline under simulated atmospheric conditions, *J. Photochem. Photobiol.*, A:Chem, 45, 195, 1988.

121. Tuazon, E.C. et al., Atmospheric reactions of N-nitrosodimethylamine and dimethylnitramine, *Environ. Sci. Technol.*, 18, 49, 1984.

TABLE 5.1 (CONTINUED)
Summary of Hazardous Air Pollutant Transformations and Lifetimes

122. Harris, G.W., Atkinson, R. and Pitts, J.N., Jr., Kinetics of the reactions of hydroxyl radical with hydrazine and methylhydrazine, *J. Phys. Chem.*, 83, 2557, 1979.

123. Harrison, R.M. and Laxen, D.P.H., Sink processes for tetraalkyllead compounds in the atmosphere, *Environ. Sci. Technol.*, 12, 1384, 1978.

124. Nielsen, O.J., Nielsen, T. and Pagsberg, P., Direct spectrokinetic investigation of the reactivity of OH with tetraalkyllead compounds in the gas phase. Estimate of lifetimes of tetraalkyllead compounds in ambient air, Riso-R-480, Riso National Laboratory, DK-4000 Roskidle, Denmark, 1984.

125. Niki, H. et al., A long-path Fourier transform infrared study of the kinetics and mechanisms for the HO-radical initiated oxidation of dimethylmercury, *J. Phys. Chem.*, 87, 4978, 1983.

126. Witte, F., Urbanik, E. and Zetsch, C., Temperature dependence of the rate constants for the addition of OH to benzene and some monosubstituted aromatics (aniline, bromobenzene and nitrobenzene) and the unimolecular decay of the adducts. Kinetics into a quasiequilibrium 2., *J. Phys. Chem.*, 90, 3251, 1986.

127. Fritz, B. et al., Laboratory kinetic investigations of the tropospheric oxidation of selected industrial emissions, In: B. Versino and H. Ott (Eds.), *Physico-Chemical Behaviour of Atmospheric Pollutants*, Reidel Publishing Co., Dordrecht, Netherlands, p. 192, 1982.

128. Atkinson, R. et al., Atmospheric chemistry of aniline, N,N-dimethylaniline, pyridine, 1,3,5-triazine and nitrobenzene, *Environ. Sci. Technol.*, 21, 64, 1987.

129. Helas, G. and Wilson, S.R., On sources and sinks of phosgene in the troposphere, *Atmos. Environ.*, 26A (16), 2975, 1992.

130. Tou, J.C. and Kallos, G.J., Kinetic study of the stabilities of chloromethyl methyl ether and bis(chloromethyl) ether in humid air, *Anal. Chem.*, 46, 1866, 1974.

131. Radding, S.B. et al., Review of environmental fate of selected chemicals, report from SRI to U.S. EPA (OTS), Contract No. 68-01-2681, EPA-560/4-75-001, 1975.

132. Morrison, R.T. and Boyd, R.N., *Organic Chemistry* (2nd ed.), Allyn and Bacon, Inc., Boston, MA, 1970.

133. Cadle, S.H., Dasch, J.M. and Mulawa, P.A., Atmospheric concentrations and the deposition velocity to snow of nitric acid, sulfur dioxide and various particulate species, *Atmos. Environ.*, 19, 1819, 1985.

134. Harrison, R.M., Rapsomanikis, S. and Turnbull, A., Land–surface exchange in a chemically reactive system; surface fluxes of HNO₃, HCl and NH₃, *Atmos. Environ.*, 23, 1795, 1989.

135. Glotfelty, D.G., Majewski, M.S. and Seiber, J.N., Distribution of several organophosphorus insecticides and their oxygen analogues in a foggy atmosphere, *Environ. Sci. Technol.*, 24, 353, 1990.

136. Hargrave, B.T. et al., Atmospheric transport of organochlorines to the arctic ocean, *Tellus*, Ser. B., 40B(5), 480, 1988.

137. Shepson, P.B. et al., Allyl chloride: The mutogenic activity of its photooxidation products, *Environ. Sci. Technol.*, 21, 568, 1987.

138. Kleindienst, T.E. et al., Generation of mutagenic transformation products during the irradiation of simulated urban atmospheres, *Environ. Sci. Technol.*, 26, 320, 1992.

139. Biermann, H.W. et al., Kinetics of the gas-phase reactions of the hydroxyl radical with naphthalene, phenanthrene and anthracene, *Environ. Sci. Technol.*, 19, 244, 1985.

140. Shepson, P.B. et al., The mutagenic activity of irradiated toluene/NOₓ/H₂O/air mixture, *Environ. Sci. Technol.*, 19, 249, 1985.

141. Dumdei, B.E. et al., MS/MS analysis of the products of toluene photoxidation and measurement of their mutagenic activity, *Environ. Sci. Technol.*, 22, 1493, 1988.

142. Holdren, M.W., Spicer, C.W. and Riggin, R.M., Gas phase reaction of toluene diisocyanate with water vapor, *Am. Ind. Hyg. Assoc. J.*, 45, 626, 1984.

143. Vigon, B.W., et al., Techniques for environmental modeling of the fate and effects of complex chemical mixtures: a case study, Aquatic Toxicology and Hazard Assessment: 10th Volume, ASTM STP 971, W.J. Adams, G.A. Chapman and W.G. Landis, Eds., American Society for Testing and Materials, Philadelphia, 247, 1988.

144. Bufalini, J.J., U.S. EPA, personal communication, 1993.

145. Atkinson, R., Gas-phase tropospheric chemistry of organic compounds: A review, *Atmos. Environ.*, 24A, 1, 1990.

146. Rogozen, M.B. et al., Evaluation of potential toxic air contaminants: Phase I, Final Report, Contract A4-131-32, CA Air Resources Board, Sacramento, CA, available from NTIS (PB 88-183330), 1987.

147. Hashimoto, S. et al., Products and mechanism for the OH radical initiated oxidation of acrylonitrile, methacrylonitrile and allylcyanide in the presence of NO, *Int. J. Chem. Kinet.*, 16, 1385, 1984.

148. Tuazon, E.C., Atkinson, R., Aschmann, S.M., Kinetics and products of the gas-phase reactions of the OH radical and O$_3$ with allyl chloride and benzyl chloride at room temperature, *Int. J. Chem. Kinet.*, 22, 981, 1990.

149. Grosjean, D., Atmospheric Chemistry of toxic contaminants: 5. Unsaturated halogenated aliphatics: ally chloride, chloroprene, hexachlorocyclopentadiene, vinylidene chloride, *J. Air Waste Manage. Assoc.*, 41, 182, 1991.

150. Edney, E.O. et al., The photooxidation of allyl chloride, *Int. J. Chem. Kinet.*, 18, 597, 1986.

151. Atkinson, R. et al., Formation of ring-retaining products from the OH radical-initiated reactions of benzene and toluene, *Int. J. Chem. Kinet.*, 21, 801, 1989.

152. Grosjean, D., Atmospheric chemistry of toxic contaminants: 1. Reaction rates and atmospheric persistence, *J. Air Waste Manage. Assoc.*, 40, 1397, 1990.

153. Atkinson, R. et al., Lifetimes and fates of toxic air contaminants in California's atmosphere, Final Report, Contract A732-107, CA Air Resources Board, Sacramento, CA, 1990.

154. Grosjean, D., Atmospheric chemistry of toxic contaminants: 2. Saturated aliphatics: acetaldehyde, dioxane, ethylene glycol ethers, propylene oxide, *J. Air Waste Manage Assoc.*, 40, 1522, 1990.

155. Harris, G.W., Atkinson, R., Pitts, J.N. Jr., Kinetics of the reactions of the OH radical with hydrazine and methylhydrazine. *J. Phys. Chem.*, 83, 2557, 1979.

156. Tuazon, E.C. et al., Products of the gas-phase reaction of methyl *tert*-butyl ether with the OH radical in the presence of NO$_x$, *Int. J. Chem. Kinet.*, 23, 1003, 1991.

157. Atkinson, R., et al., Atmospheric chemistry of aniline, N-N-dimethylaniline, pyridine, 1,3,5-triazine and nitrobenzene, *Environ. Sci., Technol.*, 21, 64, 1987.

158. Grosjean, D., Atmospheric reactions of styrenes and peroxybenzoyl nitrate, *Sci. Tot. Environ.*, 46, 41, 1985.

159. Tuazon, E.C. et al., Gas-phase reactions of 2-vinylpyridine and styrene with OH and NO$_3$ radicals and O$_3$, *Environ. Sci. Technol.*, 27, 1832, 1993.

160. Atkinson, R., Aschmann, S.M. and Arey, J., Formation of ring-retaining products from the OH radical-initiated reactions of o-, m- and p-xylene, *Int. J. Chem. Kinet.*, 23, 77, 1991.

161. Tuazon, E.C. et al., ±-dicarbonyl yields from the NO$_x$-air photooxidations of a series of aromatic hydrocarbons in air, *Environ. Sci. Technol.*, 20, 383, 1986.

162. Liu, X., Jeffries, H.E. and Sexton, K.G., Hydroxyl radical and ozone initiated photochemical reactions of 1,3-butadiene, *Atmos. Environ.*, 33, 3005, 1999.

163. Skov, H. et al., Products and mechanisms of the reactions of the nitrate radical (NO$_3$) with isoprene, 1,3-butadiene and 2,3-dimethyl-1,3-butadiene in air, *Atmos. Environ.*, 26A, 2771, 1992.

164. Yu, J., Jeffries, H.E. and Sexton, K.G., Atmospheric photooxidation of alkylbenzenes – I. Carbonyl product analysis, *Atmos. Environ.*, 31, 2261, 1997.

165. Grosjean, D., Grosjean, E. and Williams, E.L., Atmospheric chemistry of olefins: a product study of the ozone-alkene reaction with cyclohexane added to scavenge OH, *Environ. Sci. and Technol.*, 28, 186, 1994.

166. CA Air Resources Board, Toxic Air Contaminant Identification List Summaries, 1997.

167. Kwok, E.S.C. and Atkinson, R., Estimation of hydroxyl radical reaction rate constants for gas-phase organic compounds using a structure-reactivity relationship: an update, *Atmos. Environ.*, 29, 1685, 1995.

168. Atkinson, R., Gas-phase tropospheric chemistry of organic compounds, *J. Phys. Chem.* Ref. Data Monograph 2, 1-216, 1994.

169. Kwok, E.S.C., Arey, J. and Atkinson, R., Gas-phase atmospheric chemistry of dibenzo-p-dioxin and dibenzofuran, *Environ. Sci. Technol.*, 28, 528, 1994.

170. Kao, A.S., Formation and removal reactions of hazardous air pollutants, *J. Air Waste Manage. Assoc.*, 44, 683, 1994.

171. Lui, R., Huie, R.E. and Kurylo, M.J., The gas-phase reactions of hydroxyl radicals with a series of nitroalkanes over the temperature range 240-400k, *Chem. Phys. Lett.*, 167, 519, 1990.

172. Tuazon, E.C. et al., Atmospheric reactions of N-nitrosodimethylamine and dimethylamine, *Environ. Sci. Technol.*, 18, 49, 1984.

TABLE 5.1 (CONTINUED)
Summary of Hazardous Air Pollutant Transformations and Lifetimes

173. Saunders, S.M. et al., Kinetics and mechanisms of the reactions of OH with some oxygenated compounds of importance in tropospheric chemistry, *Int. J. Chem. Kin.*, 26, 113, 1994.

174. Lin, C.J. and Pehkonen, S.O., Two-phase model of mercury chemistry in the atmosphere, *Atmos. Environ.*, 32, 2543, 1998.

175. Lin, C.J. and Pehkonen, S.O., Oxidation of elemental mercury by aqueous chlorine: implications for tropospheric mercury chemistry, *J. Geophys. Res.*, 103, 28, 093, 1998.

176. Lin, C.J. and Pehkonen, S.O., The chemistry of atmospheric mercury: a review, *Atmos. Environ.*, 33, 2067, 1999.

177. Schroeder, W.H., Yarwood, G. and Niki, H., Transformation processes involving Hg species in the atmosphere—results from a literature survey, *Water, Air and Soil Pollution*, 56, 653, 1991.

178. U.S. EPA, Mercury Study Report to Congress, EPA-452/R-97-003, 1997.

179. Forstner, H.J.L., Flagan, R.C. and Seinfeld, J.H., Secondary organic aerosol from the photooxidation of aromatic hydrocarbons; molecular composition, *Environ. Sci. Technol.*, 31, 1345, 1997.

180. Leifer, A. et al., Environmental transport and transformation of polychlorinated biphenyls, Washington, D.C., U.S. EPA, Office of Toxic Substances, EPA-560/5-83-025, 1983.

181. Kwok, E.S.C, Atkinson, R. and Arey, J., Rate constants for the gas-phase reactions of the OH radical with dichlorobiphenyls, 1-chlorodibenzo-p-dioxin, e,2-dimethoxybenzene and diphenyl ether: estimation of OH radical reaction rate constants for PCBs, PCDDs and PCDFs, *Environ. Sci. Technol.*, 29, 1591, 1995.

182. Koester, C.J. and Hites, R.A., Photodegradation of polychlorinated dioxins and dibenzofurans adsorbed to fly ash, *Environ. Sci. Technol.*, 26, 502, 1992.

183. Mill, T. et al., Photolysis of tetrachlorodioxin and PCBs under atmospheric conditions, Report prepared by SRI International for USEPA, Office of Health and Environmental Assessment, Washington, D.C., 1987.

184. Orth, R.G., Ritchie, C. and Hileman, F., Measurement of the photoinduced loss of vapor phase TCDD, *Chemosphere*, 18, 1275, 1989.

185. Podoll, R.T, Jaber, H.M. and Mill, T., Tetrachlorodibenzodioxin: rates of volatilization and photolysis in the environment, *Environ. Sci. Technol.*, 20, 490, 1986.

186. Sivils, L.D. et al., Studies on gas-phase phototransformation of polychlorinated dibenzo-p-dioxins, *Organohalogen Compounds*, 19, 349, 1994.

187. Pennise, D.M. and Kamens, R.M. Atmospheric behavior of polychlorinated dibenzo-p-dioxins and dibenzofurans and the effect of combustion temperature, *Environ. Sci. Technol.*, 30, 2832, 1996.

188. Atkinson, R., Atmospheric chemistry of PCBs, PCDDs and PCDFs, *Issues in Environmental Science and Technology*, 6, 53, 1996.

189. Liu, X., Sexton, K.G. and Jeffries, H.E., Study of aromatic toxics and related pollutants in atmospheric photochemical degradation process, *Measurement of Toxic and Related Air Pollutants*, Air and Waste Management Assoc., Vol. 1, pg. 179, 1998.

190. Jang, M. and Kamens, R.M., Characterization of secondary aerosol from the photooxidation of toluene in the presence of NO_x and 1-propene, *Environ. Sci. Technol.*, 35, 3626, 2001.

Index

A

ABIOTIK$_x$, 177
Acetaldehyde, 3, 9, 14, 17, 64, 131, 134, 181, 185
Acetamide, 3, 18, 61, 64, 134, 183, 185
Acetonitrile, 3, 14, 17, 64, 134, 185
Acetophenone, 3, 14, 17, 61, 64, 134, 185
2-Acetylaminofluorene, 3, 18, 61, 64, 134, 183, 185
Acrolein, 3, 9, 14, 17, 64, 131, 134, 181, 186
Acrylamide, 3, 14, 17, 64, 135, 183, 186
Acrylic acid, 3, 14, 17, 64, 135, 183, 186
Acrylonitrile, 3, 9, 14, 17, 24, 65, 131, 135, 181, 186
Aerometric Information Retrieval System (AIRS), 126
AIRS, see Aerometric Information Retrieval System
Air and Waste Management Association (AWMA), 126
ALAPCO, see Association of Local Air Pollution Control Officers
Alcohols, 178
Aldehydes, 178
Allyl chloride, 3, 14, 17, 24, 65, 135, 186
Ambient air
    presence of HAPs in, 2
    sampling locations, distribution of HAPs by number of, 128
Ambient air, concentrations of 188 HAPs in, 125–171
    ambient air concentrations of HAPs, 127–128
    ambient air concentrations of 188 hazardous air pollutants, 134–171
    data gaps, 128–130
    recent data for high priority HAPs, 131–132
    survey procedures, 125–127
Ambient air, measurement methods for 188 hazardous air pollutants in, 55–123
    background, 56–57
    HAPs method development, 60–61
    results of survey of, 64–123
    status of current methods, 59–60
    survey approach, 57–59
Ambient concentration
    measurements, distribution of HAPs by number of, 129
    survey, intent of, 127
Ambient data archive, 126

Ambient measurement, 55, 130
American Society for Testing and Materials (ASTM), 58
4-Aminobiphenyl, 3, 18, 24, 65, 135, 183, 186
Aniline, 3, 14, 17, 24, 65, 135, 187
o-Anisidine, 3, 18, 24, 65, 135, 183, 187
Antimony compounds, 6, 18, 49, 89, 162, 213
APCI, see Atmospheric pressure chemical ionization
Aromatic compounds, 11, 129, 179
Arsenic compounds, 6, 9, 18, 49, 90, 131, 162, 181, 213
Asbestos, 3, 18, 24, 65, 135, 187
Association of Local Air Pollution Control Officers (ALAPCO), 126
ASTM, see American Society for Testing and Materials
Atmosphere, chemical and physical processes affecting HAPs in, 173
Atmospheric pressure chemical ionization (APCI), 57
AWMA, see Air and Waste Management Association
Aziridine, 4

B

Battelle Statistics and Data Analysis Systems department, 126
Benzene, 3, 9, 14, 17, 25, 65, 91, 131, 136, 181, 187
Benzidine, 3, 18, 25, 66, 136, 183, 187
p-Benzoquinone, 6, 84
Benzotrichloride, 3, 18, 25, 61, 66, 136, 187
Benzyl chloride, 3, 14, 17, 25, 66, 136, 188
Beryllium compounds, 6, 9, 18, 49, 90, 131, 162, 181, 214
Biphenyl, 3, 18, 25, 66, 136, 188
Bis(2-chloroethyl)ether, 4, 17, 17
Bis(chloromethyl)ether, 3, 14, 17, 66, 137, 188
Bis(2-ethylhexyl)phthalate, 3, 18, 25, 66, 137, 183, 188
Boiling point (BP), 12
BP, see Boiling point
Bromoform, 3, 14, 17, 66, 137, 189
Bromomethane, 5, 151
1,3-Butadiene, 3, 9, 14, 17, 66, 131, 137, 181, 189
2-Butanone, 5

C

CA, see Chemical Abstracts
CAAA, see Clean Air Act Amendments
Cadmium compounds, 6, 9, 18, 49, 90, 131, 163, 181, 214
Calcium cyanamide, 3, 18, 67, 137, 183, 189
CAP, see Chemical Abstracts Previews
Caprolactam, 3, 8, 130
Captan, 3, 18, 67, 137, 183, 189
Carbaryl, 3, 18, 67, 138, 183, 189
Carbon dioxide, 178
Carbon disulfide, 3, 14, 17, 67, 138, 189
Carbon monoxide, 178
Carbon tetrachloride, 2, 3, 9, 14, 17, 67, 131, 138, 181, 189
Carbonyl sulfide, 3, 14, 17, 67, 138, 190
Carcinogens, EPA-classified, 2
CAS, see Chemical Abstracts Service
Catechol, 3, 14, 17, 67, 138, 183, 190
CERCLA, see Comprehensive Emergency Response and Compensation Liability Act
Chemical Abstracts (CA), 125, 177
Chemical Abstracts Previews (CAP), 125, 177
Chemical Abstracts Service (CAS), 12, 59
Chloramben, 3, 18, 61, 67, 139, 183, 190
Chlordane, 3, 18, 68, 139, 183, 190
Chlorinated camphene, 46, 87
Chlorine, 3, 18, 68, 139, 190
Chloroacetic acid, 3, 14, 17, 68, 139, 190
2-Chloroacetophenone, 3, 18, 68, 139, 191
Chlorobenzene, 3, 14, 17, 68, 139, 191
Chlorobenzilate, 3, 18, 69, 139, 183, 191
2-Chlorobiphenyl, 42
2-Chloro-1,3-butadiene, 28, 69, 140, 191
1-Chloro-2,3-epoxy propane, 4, 33, 197
Chloroethane, 4
Chloroform, 2, 3, 9, 14, 17, 28, 69, 131, 140, 181, 191
Chloromethane, 5, 151
Chloromethyl methyl ether, 3, 14, 17, 28, 69, 140, 191
Chloroprene, 3, 14, 17, 28, 69, 140, 191
Chromium compounds, 6, 9, 18, 49, 90, 131, 163, 181, 214
Clean Air Act, 174
Clean Air Act Amendments (CAAA), 1, 55
CLP, see U.S. Environmental Protection Agency Contract Laboratory Program
Cobalt compounds, 6, 18, 49, 90, 163, 214
Coke oven emissions, 6, 9, 11, 18, 49, 91, 131, 163, 181, 183, 214
Compound class, data completeness by, 179
Comprehensive Emergency Response and Compensation Liability Act (CERCLA), 2

Contract required quantitation limits (CRQL), 59
Coronene, 50, 51, 91, 92
Cresol, 3, 17, 28, 69, 126, 140, 191
m-Cresol, 3, 18, 28, 69, 140, 192
o-Cresol, 3, 14, 17, 28, 69, 140, 192
p-Cresol, 3, 18, 70, 140, 192
Cresylic acid, 3, 14, 17, 28, 69, 140, 191
Criteria pollutant, 2
CRQL, see Contract required quantitation limits
Cumene, 3, 14, 17, 70, 141, 192
Cyanide
    compounds, 6, 18, 50, 91, 163, 214
    particulate, 91

D

2,4-D, see 2,4-Dichloro phenoxyacetic acid, 18
DDE, see 1,1-Dichloro-2,2-bis(p-chlorophenyl)ethylene
Diazomethane, 4, 14, 17, 71, 142, 183, 193
Dibenzofuran, 4, 18, 71, 142, 193
1,2-Dibromo-3-chloropropane, 4, 14, 17, 30, 71, 143, 193
Dibromoethane, 4
1,2-Dibromoethane, 9, 34
Dibutylphthalate, 4, 18, 30, 71, 143, 183, 193
1,4-Dichlorobenzene, 4, 14, 17, 30, 71, 143, 194
3,3'-Dichlorobenzidine, 4, 18, 30, 72, 143, 183, 194
1,1-Dichloro-2,2-bis(p-chlorophenyl)ethylene (DDE), 4, 18, 70, 142, 183, 193
1,1-Dichloroethane, 4
1,2-Dichloroethane, 4, 9, 34, 35
1,1-Dichloroethylene, 6
Dichloroethyl ether, 4, 14, 17, 30, 72, 143, 194
Dichloromethane, 5, 80
2,4-Dichlorophenoxyacetic acid (2,4-D), 3, 18, 70, 193
1,2-Dichloropropane, 5, 9, 84, 156
1,3-Dichloropropene, 4, 9, 14, 17, 31, 72, 131, 132, 143, 181, 194
Dichlorvos, 4, 18, 31, 72, 144, 183, 194
Diethanolamine, 4, 18, 31, 72, 144, 183, 195
1,4-Diethyleneoxide, 4, 33, 197
Diethyl sulfate, 4, 14, 17, 31, 72, 144, 195
3,3'-Dimethoxybenzidine, 4, 18, 31, 73, 144, 183
Dimethylaminoazobenzene, 183
4-Dimethylaminoazobenzene, 4, 18, 31, 73, 144, 195
N,N-Dimethylanaline, 4, 14, 17, 31, 73, 144, 195
3,3'-Dimethylbenzidine, 4, 18, 31, 73, 144, 183, 195
Dimethylcarbamoyl chloride, 4, 14, 17, 31, 73, 144, 183, 195
N,N-Dimethylformamide, 4, 14, 17, 32, 73, 145, 183, 195
1,1-Dimethylhydrazine, 4, 14, 17, 32, 73, 145, 196
Dimethyl phthalate, 4, 18, 32, 73, 145, 196

Dimethyl sulfate, 4, 14, 17, 32, 73, 145, 196
4,6-Dinitro-o-cresol, 4, 18, 32, 74, 145, 196
2,4-Dinitrophenol, 4, 18, 33, 74, 145, 197
2,4-Dinitrotoluene, 4, 18, 33, 74, 145, 197
1,4-Dioxane, 4, 14, 17, 33, 74, 145, 197
1,2-Diphenylhydrazine, 4, 18, 33, 61, 74, 145, 183, 197
Direct air sampling mass spectrometry, 57
Dispersion modeling, 55
Dye intermediates, 130

E

Electronic polarizability, 20
Emergency Planning and Community Right to Know Act of 1986 (EPCRA), 2
Emission measurements, 62
EPA, see U.S. Environmental Protection Agency
EPCRA, see Emergency Planning and Community Right to Know Act of 1986
Epichlorohydrin, 4, 14, 17, 33, 74, 145, 197
1,2-Epoxybutane, 4, 14, 17, 34, 74, 145, 197
Ethyl acrylate, 4, 14, 17, 34, 74, 146, 198
Ethylbenzene, 4, 14, 17, 34, 75, 146, 198
Ethyl carbamate, 4, 15, 17, 34, 61, 75, 146, 183, 198
Ethyl chloride, 4, 15, 17, 34, 75, 146, 198
Ethylene dibromide, 4, 9, 15, 17, 34, 75, 131, 146, 181, 198
Ethylene dichloride, 4, 9, 15, 17, 34, 75, 131, 147, 181, 199
Ethylene glycol, 4, 18, 35, 75, 147, 199
Ethyleneimine, 4, 15, 17, 35, 75, 147, 183, 183, 199
Ethylene oxide, 4, 9, 15, 17, 35, 75, 131, 147, 181, 199
Ethylene thiourea, 4, 18, 35, 76, 199
Ethylidene dichloride, 4, 15, 17, 35, 76, 147, 199

F

Fine mineral fibers, 6, 18, 51, 92, 165, 215
Formaldehyde, 2, 4, 9, 15, 17, 35, 76, 131, 147, 181, 199
Fourier-transform infrared spectroscopy (FTIR), 57
FTIR, see Fourier-transform infrared spectroscopy

G

Gas phase HAPs, 173
Glycol ethers, 6, 18, 50, 91, 163, 214

H

Halogenated aromatics, 11, 179
Halogenated hydrocarbons, 129, 179

Halogen-containing, hydrocarbons, 11
HAP, see Hazardous air pollutant
Hazardous air pollutant (HAP), 1
    ambient air concentrations of, 127
    ambient air measurement methods for, 64–123
    categories, summary of with defined vapor pressure ranges, relevant properties, and number of HAPs in each category, 13
    common features of Title III, 11
    emission rates, 55
    gas phase, 173
    high priority, 180
        lifetime and transformation data for, 181
        recent data for, 131–132
        urban, 64
    history of definition of, 1
    list, impact of, 8
    measurements, source-related, 55
    method development, 60
    number of for each chemical category, 129
    presence of in ambient air, 2
    transformation(s), 10, 177
        example of large indoor environmental chamber used to study, 176
        outdoor Teflon chamber used to study, 177
        small laboratory chamber used to study, 176
    use of vapor pressure data to categorize, 10
    vapor pressure of, 56
Hazardous air pollutants, introduction to, 1–10
    background, 1–2
    hazardous air pollutants under Clean Air Act Section 112(b), 3–7
    impact of HAPs list, 8–9
    list of hazardous air pollutants, 2–8
    organization of information, 9–10
Heptachlor, 4, 18, 35, 76, 148, 183, 199
Hexachlorobenzene, 4, 9, 18, 36, 76, 131, 132, 148, 181, 200
Hexachlorobutadiene, 4, 15, 17, 36, 76, 149, 200
1,2,3,4,5,6-Hexachlorocyclohexane, 4, 18, 36, 77, 149, 200
Hexachlorocyclopentadiene, 4, 18, 36, 77, 149, 200
Hexachloroethane, 4, 15, 17, 36, 77, 149, 200
Hexamethylene diisocyanate, 4, 18, 77, 149
Hexamethylene-1,6-diisocyanate, 36, 200
Hexamethylphosphoramide, 5, 18, 37, 61, 77, 149, 183, 201
Hexane, 5, 15, 17, 37, 77, 149, 201
High priority HAPs
    lifetime and transformation data for, 181
    recent data for, 132
High priority urban air toxics, HAPs designated as, 131
Hydrazine, 5, 9, 18, 37, 77, 131, 149, 181, 201
Hydrocarbons, 11, 129, 179

halogen-containing, 11
nitrogen-containing, 11
oxygen-containing, 11
Hydrochloric acid, 5, 18, 77, 201
Hydrofluoric acid, 5, 18, 78
Hydrogen chloride, 5, 18, 37, 150
Hydrogen cyanide, 50, 91
Hydrogen fluoride, 5, 18, 37, 78, 150, 201
Hydroquinone, 5, 18, 37, 78, 150, 201

I

Inorganics, 11, 56, 129, 179
Integrated Urban Air Toxics Strategy (IUATS), 8
Iodomethane, 5
Isophorone, 5, 15, 17, 37, 78, 150, 183, 201
IUATS, see Integrated Urban Air Toxics Strategy

K

Ketones, 178
Kinetics
    selection of reaction vessel for, 175
    studies, 174

L

Lead compounds, 6, 9, 18, 50, 91, 131, 164, 181, 215
Lindane, 183
Long-path optical methods, 57

M

MACT, see Maximum achievable control technology
Maleic anhydride, 5, 18, 37, 78, 150, 201
Manganese compounds, 6, 9, 18, 51, 91, 131, 164, 181, 215
Mass spectrometry (MS), 57
Maximum achievable control technology (MACT), 1
Mercury compounds, 6, 9, 18, 51, 92, 131, 132, 165, 181, 215
Methanol, 5, 15, 17, 38, 78, 150, 202
Methoxychlor, 5, 18, 38, 78, 151, 183, 202
2-Methylaziridine, 6, 61, 84
Methyl bromide, 5, 15, 17, 38, 79, 151, 202
Methyl tert-butyl ether, 5, 15, 17, 39, 80, 153, 203
Methyl chloride, 5, 15, 17, 38, 79, 151, 202
Methyl chloroform, 5, 15, 17, 38, 79, 151, 202
4,4'-Methylenebis-(2-chloroaniline), 5, 18, 39, 80, 153, 183, 203
Methylene chloride, 5, 9, 15, 17, 39, 80, 131, 153, 181, 203
4,4'-Methylenedianiline, 5, 18, 39, 81, 153, 183, 203

4,4'-Methylenediphenyl diisocyanate, 5, 18, 39, 80, 153, 203
Methyl ethyl ketone, 5, 15, 17, 38, 79, 152, 202
Methylhydrazine, 5, 15, 17, 38, 79, 153, 203
Methyl iodide, 5, 15, 17, 38, 79, 153, 203
Methyl isobutyl ketone, 5, 15, 17, 38, 79, 153, 203
Methyl isocyanate, 5, 15, 17, 39, 79, 152, 183, 203
Methylmethacrylate, 5, 15, 17, 39, 80, 152, 203
Molar reflectivity, 13
Molecular weight (MW), 12
MS, see Mass spectrometry
Mutagenic activity, 182
MW, see Molecular weight

N

Naphthalene, 5, 18, 39, 50, 51, 81, 91, 92, 153, 204
NATICH, see EPA National Air Toxics
    Information Clearinghouse
National Institute of Occupational Safety and Health
    (NIOSH), 58, 59
National Technical Information Service (NTIS), 59, 125, 177
Nickel compounds, 6, 9, 18, 51, 92, 131, 166, 181, 215
NIOSH, see National Institute of Occupational
    Safety and Health
Nitrates, 178
Nitrobenzene, 5, 15, 17, 39, 81, 154, 204
4-Nitrobiphenyl, 5, 39, 81, 154, 204
Nitrogenated organics, 129, 179
Nitrogen-containing hydrocarbons, 11
4-Nitrophenol, 5, 18, 40, 81, 154, 204
2-Nitropropane, 5, 15, 17, 40, 81, 154, 204
N-Nitrosodimethylamine, 5, 15, 17, 40, 82, 154, 205
N-Nitroso-N-methylurea, 5, 15, 17, 40, 61, 81, 154, 204
N-Nitrosomorpholine, 5, 15, 17, 40, 82, 154, 205
Nonvolatile organic compounds (NVOCs), 12, 125
NTIS, see National Technical Information Service
NVOCs, see Nonvolatile organic compounds

O

OAQPS, see EPA Office of Air Quality Planning and
    Standards
Occupational Safety and Health Administration
    (OSHA), 58, 59
Organic acids, 178
Organic compounds, multifunctional, 178
Organics
    nitrogenated, 129
    semivolatile, 56
    volatile, 56

OSHA, see Occupational Safety and Health
     Administration
Oxygenated organics, 179
Oxygen-containing hydrocarbons, 11

P

Parathion, 5, 18, 41, 82, 154, 205
Particulate
     cyanide, 91
     -phase inorganics, 56
PCBs, see Polychlorinated biphenyls
Pentachloronitrobenzene, 5, 18, 41, 82, 155, 205
Pentachlorophenol, 5, 18, 41, 82, 155, 205
Pesticides, 11, 129, 179
Phenol, 5, 15, 17, 41, 82, 155, 205
p-Phenylenediamine, 5, 18, 41, 83, 155, 183, 206
Phosgene, 5, 15, 17, 42, 83, 155, 206
Phosphine, 5, 18, 42, 83, 155, 206
Phosphorus, 5, 18, 42, 83, 155, 206
Photolysis, 173
Phthalates, 11, 129, 179
Phthalic anhydride, 5, 18, 42, 83, 155, 206
Physical Properties Database (PHYSPROP), 16
PHYSPROP, see Physical Properties Database
Pollutant, see also Hazardous air pollutant
     criteria, 2
     exposure distributions, 56
Polychlorinated biphenyls (PCBs), 2, 5, 9, 11, 18,
     42, 83, 131, 155, 181, 206
Polycyclic aromatic compounds, 130
Polycyclic aromatic hydrocarbons, 125
Polycyclic organic matter (POM), 6, 9, 18, 51, 92,
     126, 131, 166, 181, 215
POM, see Polycyclic organic matter
Product studies, 174
1,3-Propane sultone, 5, 15, 17, 43, 83, 156, 207
Propene, atmospheric transformation of, 182
β-Propiolactone, 5, 15, 17, 43, 83, 156, 207
Propionaldehyde, 5, 15, 17, 43, 83, 156, 207
Propoxur, 5, 18, 43, 84, 183, 156, 207
Propylene dichloride, 5, 9, 15, 17, 43, 84, 131, 156,
     180, 181, 207
1,2-Propyleneimine, 6, 15, 17, 43, 61, 84, 157, 183,
     207
Propylene oxide, 5, 15, 17, 43, 84, 156, 207

Q

Quinoline, 6, 9, 18, 44, 84, 131, 157, 180, 181, 207
Quinone, 6, 18, 44, 84, 157, 208
Quintobenzene, 5, 82

R

Radionuclides, 6, 8, 52, 92, 128, 167, 179, 216
Radon, 6, 18, 52, 92, 167, 179, 216
Reaction kinetics, experimental investigation of, 175
Reaction products, studies designed to measure, 174

S

Selenium compounds, 6, 18, 52, 93, 167, 216
Semivolatile organic compounds (SVOCs), 11, 12,
     56, 125
Stack emission measurements, 62
STAPPA, see State and Territorial Air Pollutant
     Program Administrators
State and Territorial Air Pollutant Program
     Administrators (STAPPA), 126
STN International, 177
Styrene, 6, 15, 17, 44, 85, 157, 208
Styrene oxide, 6, 15, 17, 44, 85, 157, 208
Sulfates, 11, 129, 179
Superfund, see Comprehensive Emergency
     Response and Compensation Liability
     Act
SVOCs, see Semivolatile organic compounds

T

2,3,7,8-TCDD, 9, 131, 132
2,3,7,8-Tetrachlorodibenzo-p-dioxin, 6, 18, 44, 85,
     157, 181, 208
1,1,2,2-Tetrachloroethane, 6, 9, 15, 17, 44, 85, 131,
     158, 181, 208
Tetrachloroethylene, 6, 9, 15, 17, 45, 85, 131, 158,
     181, 209
Titanium tetrachloride, 6, 18, 45, 61, 85, 130, 158,
     209
Title III hazardous air pollutants, 11–53
     atmospheric transformation products of Clean Air
          Act, 173–224
     experimental approaches for study of HAP
          transformations, 174–176
     hazardous air pollutant transformations,
          177–180
     summary of hazardous air pollutant
          transformations and lifetimes, 185–224
     transformations of other atmospheric
          chemicals, 181–182
     transformations of 33 HAPs, 180
     chemical and physical properties of 188 HAPs,
          12–13
     common features of, 11–12
     HAPs grouped by volatility class, 17–19
     188 hazardous air pollutants, 11

polarizability and water solubility as defining
    characteristics of polar and nonpolar
    VOCs, 13–20
properties, sources/uses, and chemical/volatility
    group classifications of CAAA Title III
    HAPs, 23–53
Toluene, 6, 15, 17, 45, 86, 91, 158, 209
Toluene-2,4-diamine, 6, 18, 86, 209
2,4-Toluenediamine, 45, 159, 183
Toluene 2,4-diisocyanate, 159
2,4-Toluene diisocyanate, 6, 18, 45, 86, 209
o-Toluidine, 6, 18, 45, 86, 159, 183, 210
Toxaphene, 6, 18, 46, 87, 159, 210
Transformation products, 178
1,2,4-Trichlorobenzene, 6, 15, 17, 46, 87, 159,210
1,1,1-Trichloroethane, 5, 151
1,1,2-Trichloroethane, 6, 15, 17, 46, 87, 159, 210
Trichloroethylene, 6, 9, 15, 17, 46, 87, 131, 160, 181,
    211
2,4,5-Trichlorophenol, 6, 18, 47, 87, 160, 211
2,4,6-Trichlorophenol, 6, 18, 43, 47, 87, 160, 211
Triethylamine, 6, 17, 47, 88, 160, 211
Trifluralin, 6, 18, 47, 88, 160, 211
2,2,4-Trimethylpentane, 6, 17, 47, 88, 160, 211

U

Urethane, 4, 75
U.S. Environmental Protection Agency (EPA), 1
    /Air and Waste Management Association, 126
    -classified carcinogens, 2
    Compendium IO-Methods, 58
    Compendium Method TO-14, 129
    Compendium Method TO-14A, 58
    Contract Laboratory Program (CLP), 58
    National Air Toxics Information Clearinghouse
        (NATICH), 7

Office of Air Quality Planning and Standards
    (OAQPS), 126
Screening Methods, 58, 59

V

Vapor pressure (VP), 10, 12
Variable volume vessels, 175
Very volatile inorganic compounds (VVINC), 12
Very volatile organic compounds (VVOC), 12
Vinyl acetate, 6, 17, 47, 88, 161, 212
Vinyl bromide, 6, 17, 47, 88, 161, 212
Vinyl chloride, 6, 9, 17, 47, 89, 131, 161, 181, 212
Vinylidene chloride, 6, 17, 47, 89, 161, 212
VOCs, see Volatile organic compounds
Volatile organic compounds (VOCs), 11, 56, 125
    classifying, 16
    nonpolar, 13, 16
    polar, 13, 16
Volatility class, 12
VP, see Vapor pressure
VVINC, see Very volatile inorganic compounds
VVOC, see Very volatile organic compounds

W

Water solubility, 20
Workplace emission measurements, 62

X

Xylene, 6, 17, 48, 89, 91, 126, 161, 212
m-Xylene, 6, 17, 48, 89, 162, 213
o-Xylene, 6, 17, 48, 89, 161, 213
p-Xylene, 6, 17, 48, 89, 162, 213